D1523810

Perspectives on Rhetorical Invention

Perspectives on Rhetorical Invention

Edited by

Janet M. Atwill
and
Janice M. Lauer

Tennessee Studies in Literature Volume 39

The University of Tennessee Press Knoxville

TENNESSEE STUDIES IN LITERATURE

"Tennessee Studies in Literature," a distinguished series sponsored by the Department of English at The University of Tennessee, Knoxville, began publication in 1956. Beginning in 1984, with Volume 27, TSL evolved from a series of annual volumes of miscellaneous essays to a series of occasional volumes, each one dealing with a specific theme, period, or genre, for which the editor of that volume has invited contributions from leading scholars in the field.

First Edition.
This book is printed on acid-free paper.

LIBRARY OF CONGRESS CATALOGING-IN-PUBLICATION DATA

Perspectives on rhetorical invention / edited by Janet M. Atwill and Janice M. Lauer.— 1st ed.
p. cm. — (Tennessee studies in literature; v. 39)
Includes bibliographical references and index.
ISBN 1-57233-201-8 (cl.: alk. paper)
1. Invention (Rhetoric) I. Atwill, Janet M.
II. Lauer, Janice M. III. Series.
P301.5.I57 P47 2002
808—dc21 2002004366

In loving memory of

David

Contents

Foreword

JANICE M. LAUER

In the mid-twentieth century, discussions of invention were inextricably woven with attempts to revive an interest in rhetoric within the academy, particularly in English studies. Elbert Harrington articulated this connection in his 1962 essay, "A Modern Approach to Invention," in the *Quarterly Journal of Speech,* contending that "most teachers know that rhetoric has always lost life and respect to the degree that invention has not had a significant and meaningful role" (373). Through complex historical circumstances, rhetorical invention had been either folded into logic (Ramus), limited to finding the known (Bacon), banished altogether from rhetoric (Blair), or devoted to "proving the truth" (Hope). The dormant if not dead state of rhetoric could be seen in the power structures of the academy in which literature eclipsed rhetoric and philosophy controlled invention. With rhetoric's loss of life and respect came the loss of power. In the early twentieth century, philosophy held sway over the study of reasoning of all kinds, restricting it to formal logic, even symbolic logic. English studies held sovereignty over written discourse, focusing only on literary discourse, abandoning rhetoric as a discipline and keeping only its application—the teaching of composition. Within composition teaching, invention was neglected, contributing to the loss of prestige and power of composition instructors.

Remarkably, within the next three decades, a plethora of inventional studies emerged within different fields in the academy, intertwined with other connected movements. In the sixties, the revival of interdisciplinary interest in rhetoric (e.g., The Rhetoric Society of America) contributed to the development of the discipline of Rhetoric and Composition and to this surge of research on invention in English studies. In 1964, I began investigating the state of invention in both contemporary theory and composition pedagogy. By the time the study was completed, I had found new and diverse work on invention, including studies by Perelman and Olbrechts-Tyteca and Toulmin on informal argument, adaptations of Burke's Pentad, Scott's writing on rhetoric as epistemic, Corbett's discussions on classical invention, Young, Becker, and Pike's tagmemic rhetoric as a process of inquiry, and Gordon and Wlecke's prewriting research. Entailed in these studies was a challenge to the domination of formal reasoning and an examination of the nature of inventional thinking. I argued for the relevance of studies on heuristic thinking as descriptive of the open-ended yet guided nature of inventional acts in composing written discourse. Janet Emig's study of composing processes theorized a frame for inventional acts. This work on invention in the sixties was followed in the seventies by other theories such as D'Angelo's conceptual theory of discourse, and Flower and Hayes's cognitive process model. In the eighties, Karen LeFevre argued for invention as a social act, prompting meta-theoretical discussions of prior theories, and Flower constructed a socio-cognitive theory of writing, describing collaborative planning. Throughout these decades, classrooms and textbooks continued to advocate inventional practices.

During the nineties, however, interest in invention appeared to wane in the field of Rhetoric and Composition. The purpose of this collection of essays, therefore, is to offer a forum for continued work on invention within the framework of recent developments in postmodernism, revisionist historiography, cultural studies, writing in the disciplines, technology, and other areas. If inventional research is to continue to flourish, it must remain sensitive to shifts in epistemology, ethics, and politics. The essays in this volume undertake this effort.

Acknowledgments

Special thanks are due Susan North for research assistance and D. Allen Carroll and Don Richard Cox for their support of this project.

Introduction

Finding a Home or Making a Path

JANET M. ATWILL

The idea for this collection grew out of formal and informal discussions on the status of research in rhetorical invention at the beginning of the millennium. Janice Lauer and I had both expressed frustration over what we perceived as a neglect of the rhetorical canon of invention, the canon that was most responsible for the renaissance of rhetorical studies in the last half of the twentieth century. She did her own research, leading to the first essay in this collection, "Rhetorical Invention: The Diaspora." Her survey of the field brought her to two tentative conclusions. First, research in invention was being conducted, but it had, in her words, "migrated, entered, settled, and shaped" other research areas in rhetoric and composition, such as writing across the disciplines and cultural studies. Second, her investigation suggested that this research tended to be more focused on theory than practice. In other words, while researchers theorized about the concept of inquiry, they were less likely to craft the kind of heuristics generally considered to be the fullest expression of the canon.

The essays in this collection both confirm and challenge these conclusions. But the conclusions themselves raised a number of questions. Why does this research tend to privilege theory over practice? Why does the metaphor of migration seem such an apt description of invention? Is there something in the character of invention that makes it prone to migrate? Or is there something in the institution that makes invention ill at ease?

When Aristotle defined rhetoric as the art of observing the available means of persuasion, he placed the art in a peculiar place between theory and practice, subjectivism and empiricism, the aesthetic and the utilitarian. These binary oppositions have never served invention very well. Indeed, the status of the art would seem to be fairly described by Jacques Derrida in *The Truth in Painting* as "a place that is neither theoretical nor practical or else both theoretical and practical" (38). While this indeterminacy has productive potential, Derrida also points to the risk of inhabiting such an ill-defined space: "But this here, this place is announced as a place deprived of place. It runs the risk, in taking place, of not having its own proper domain" (38).

We might agree with Derrida that art's productive capacity lies in its potential to subvert rather than secure binary oppositions. However, the history of research in invention in the last half of the twentieth century—at least in American higher education—suggests that institutions are ill designed to accommodate this indeterminacy. When confronted with these oppositions, the institution has shown a propensity to choose theory over practice and to accommodate the subjective and aesthetic over the empirical and utilitarian. To be sure, the complex character of rhetorical invention is responsible in part for its ambiguous status. Invention is concerned with practice, but it aims at creating arts that can inform practice across situations. Moreover, while the art aims at enabling practice, throughout its history it has been defended as being more than an instrumental means to an end. Still, while historical research in invention has continued relatively uninterrupted in classical studies and speech communication, other forms of research in invention have faced a number of challenges, especially research in English studies.

Early studies in invention were conceived as "useful" responses to the needs of composition, but this research provoked controversy as it touched on a number of closely related oppositions: humanities set against technology; creativity set against problem solving; the individual set against society. Research in the sixties and seventies was interdisciplinary, frequently drawing on linguistics, but some of that research also drew on empirical studies and methodologies from the social sciences. Most early researchers were careful to link their work to literary studies or the humanistic tradition—as if anticipating that exploring this canon that was so concerned with practical activity would be met with questions. For example, Richard Young and Alton Becker opened their 1965 article, "Toward a Modern Theory of Rhetoric: A Tagmemic Contribution," with the statement that "the heart of a liberal education was the trivium of grammar, logic, and rhetoric"; and "modern linguistics," they asserted, had "come to encompass more and more of this trivium" (135). Ross Winterowd similarly allied his research with linguistics, drawing in particular on Roman Jakobson's schema analysis. However, in the introduction to his *Contemporary Rhetoric,* he used this linguistic method to analyze the poetry of Allen Ginsberg. Janet Emig unabashedly advocated empirical methodologies for studying writing. Still, in situating *The Composing Process of Twelfth Graders,* she compared and contrasted her case study method to the first-person accounts of writing by such authors as James Joyce, Thomas Mann, and William Wordsworth.

Janice Lauer's well-known debate with Ann Berthoff foregrounded perceptions of what was at stake in transgressing the boundary between the humanities and the social sciences. Less apologetic than her colleagues, Lauer insisted in her 1970 *College Composition and Communication* article, "Heuristics and Composition," that it was time for writing teachers to "break out of the ghetto" and "investigate beyond the field of English, beyond even the area of rhetorical studies" (396). She proposed that compositionists explore research on heuristics in creative problem solving being conducted in psychology, and her article included an extensive bibliography to encourage exploration.

In "The Problem of Problem Solving," published in *College Composition and Communication* in 1971, Berthoff attacked Lauer's proposal on several fronts. She argued that Lauer set problem solving in opposition to creativity, despite the fact that Lauer had consistently used the term "creative problem solving." Most of Berthoff's critique, however, was mounted on disciplinary, institutional, and political grounds. Lauer was faulted for using the research of psychologists, whom Berthoff referred to as "technologists of learning" whose approaches were "politically not above suspicion" (237). She accused Lauer of failing to consider the "crucial interdependency of psychological and political factors" (237). Berthoff insisted that a "psychology of learning, no matter how carefully researched or how liberal its assumptions, can be politically dangerous unless it is conceived in the context of a sound sociology of knowledge" (239). For Berthoff, a "sound sociology of knowledge" required defining "the common ground of all school work, of all disciplines" (240). Moreover, this undertaking was, according to Berthoff, "philosophical precisely because it is concerned with that juncture of the public and personal, the social and individual, the political and the psychological" (240). Berthoff concluded that the questions raised by Lauer (and presumably other composition researchers like her) could be adequately answered without stepping outside the humanities building.

What was missing from Berthoff's critique was a sense of what classical philosophers called the domain of probable knowledge. It was in this domain that Aristotle placed not only rhetoric but also the political and ethical knowledge that informed public, civic discourse. In the institution depicted by Berthoff, there seemed to be only two types of knowledge: the humanistic exploration of value, on the one hand, and hard, instrumental, scientific knowledge, on the other. Lauer touched on this distinction in her response to Berthoff's article, pointing out that Berthoff was collapsing science and technology. Lauer argued that though researchers used empirical, "scientific" methodologies, they did not view creative problem solving as a closed, rule-bound process ("Response to Berthoff" 209).

Berthoff's second response, "Counterstatement," seemed to confirm that what was finally at stake were disciplinary boundaries and the character and status of the humanities. Berthoff insisted that the questions to which Lauer turned to cognitive psychologists for answers were questions that "we can ask pretty much on our own" (415). The answers could be found "by looking again at those writers we happen to admire; by reading the notebooks and journals of artists and thinkers . . . and by talking with present-day artists and artisans" (415). Thus Berthoff concluded: "If we make use of the knowledge we have as teachers of English, we can pursue such speculation fruitfully, without the guidance of psychologists who are studying the 'area' of 'creativity.' For creativity is not an area; it is the heart of the matter and the matter is using the mind to create images and models by means of language" (415).

It would be unfair to separate the Lauer-Berthoff exchange from its historical context. Berthoff was quite accurate in her assessment of the encroaching influence on the university of science and technology, an influence funded in large part by defense research contracts. The dehumanized uses of scientific and technological research had been critiqued in the work of the Frankfurt School and C. Wright Mills. Thus, early research in invention was conducted in a time of institutional change and political turmoil. Moreover, it is not surprising that traditional humanists might have been uncomfortable with the research methodologies of other fields. But, for Berthoff, the question of research methodology became a question of politics. It is difficult not to believe that such reactions had a chilling effect on research in invention.

The issue of methodology and politics in research in invention was raised again in the late eighties in James Berlin's critiques of cognitive rhetoric. Berlin's critique was explicitly leftist, and his postmodern orientation raised some elements of the debate to a higher power. In "Rhetoric and Ideology in the Writing Class," Berlin challenged the work of Linda Flower and John Hayes for its empirical orientation. According to Berlin, this orientation was based on a naive epistemological foundationalism, which, Berlin argued, betrayed its compatibility—if not complicity—with the "new American university system," whose primarily mission was to rationalize and enable competitive capitalism (480). As Berlin used them, the terms "science" and "empirical" were so charged as to function themselves as indictments. Berlin observed that "there is no question that Flower considers her work to fall within the domain of science"; indeed, "her statements about the composing process of writing . . . are based on empirical findings, on 'data-based' study, specifically the analysis of protocols recording the writing of choices of both experienced and inexperienced writers" (481). Berlin argued that the cognitive paradigm's view of language suggests that there is "a beneficent correspondence between the structures of the mind, the structures of the world, the structures of the minds of the audience, and the structures of language"(483). Flower's use of the terms "problem solving" and "goal directed" were cited as evidence that "the rationalization of the writing process is specifically designated an extension of the rationalization of economic activity" (483). The cognitive paradigm was further suspect because it focused on the writer as an individual: "problem solving is finally the act of an individual performing in isolation, solitary and alone" (482). Thus Berlin concluded that cognitive rhetoric was "eminently suited to appropriation by the proponents of a particular ideological stance, a stance consistent with the modern college's commitment to preparing students for the world of corporate capitalism" (482).

Like Lauer, Flower argued that Berlin's critique misconstrued the character of her research. In her response to Berlin, Flower defended the probabilistic nature of such inquiry, explaining: "We build theories and models in order to test our perceptions against experience, even though such 'tests' must still rely on a theoretical perspective and probabilistic claims, whether they are based on rhetoric or statistics" (766). However, once more research methodology was tied to politics, and this version of invention research was characterized as epistemologically and politically suspect.

History would moderate both sets of exchanges. Both Berthoff and Lauer elaborated on issues raised in shorthand in their discussion in *College Composition and Communication.* Berlin offered a more nuanced understanding of institutional culture in his last book, *Rhetoric, Poetics, and Cultures*. Flower applied the problem-solving model to groups and community issues, as her article with Julia Deems in this collection details. Still, it is difficult not to believe that the intensity of these critiques did not shape the direction of research in invention, inhibiting the kind of cumulative research that helps a field of study develop and mature.

As postmodern theory gained influence in humanistic studies, one might have expected its deconstruction of binary oppositions to have made the canon of invention more comprehensible. More often, however, postmodernism has been deployed to challenge invention. For example, in "Antifoundationalism, Theory Hope, and the Teaching of Composition," Stanley Fish argued that the knowledge associated with practice is so situation-bound that theories can never cross situations to shape and inform practice. Indeed, since theory-making, itself, is its own discrete practice, theory can have "no consequences." Fish's conception of postmodernism would seem to offer little to inform research in invention since the very purpose of inventional strategies is to enable practice across rhetorical situations. Other versions of postmodernism have significantly more to offer to our understanding of invention. Pierre Bourdieu's theory/practice critiques, for example, hold potential for elucidating both commonsense, probable knowledge and the kind of knowledge called "art." For the most part, however, invention has been put on the defensive by postmodernism.

The essays in this collection may then be viewed as both legacies and interventions into the institutional values and state of research in invention. They are legacies in that they continue to grapple with the opposition between theory and practice. Moreover, while these essays are interdisciplinary, only a few invoke methodologies outside the humanities. They are interventions on several points. Several of these essays confront the relationship between invention and postmodernism—some by refiguring invention, others by challenging postmodernism. Others examine invention in light of rhetoric's civic function, seeking to redefine that function for our own postmodern polis. Scholars also offer cultural and historical perspectives that enlarge our conception of invention.

Debra Hawhee's "Kairotic Encounters" uses concepts from classical rhetoric and Gilles Deleuze and Félix Guattari to outline a postmodern theory of invention. She argues that modernism has constrained our understanding of invention by two binary conceptions of invention: invention as the discovery of a pre-existent object or the creative production of a unified subject. According to Hawhee, both conceptions of invention depend on a traditional notion of subjectivity. Hawhee offers in its place a conception of subjectivity and invention that she describes as "invention-in-the-middle." She draws on several sources to outline this notion of invention: the middle voice in Greek; the concepts of *kairos, intermezzo,* and *dunamis;* and Gorgias' style in the *Helen.* According to this conception of invention, the subject is the outcome rather than the source of the rhetorical situation, fluidly acting in the moment to effect change.

In "Rhetoric and Hermeneutics: Division Through the Concept of Invention," Arabella Lyon addresses the institutional values that have suppressed rhetoric's public function by privileging interpretation over invention—hermeneutics over heuristics. Lyon holds postmodernism's concern with textuality partly to blame, maintaining that this turn "toward interpretation and away from production and 'making'" has effaced rhetoric's public function—or, as she puts it, rhetoric's identification with "action in the world." Lyon examines the hermeneutic/heuristic opposition as it appears in Michael Leff's debates with Dilip Gaonkar and in the work of Steven Mailloux. She concludes that an adequate understanding of the hermeneutic/heuristic distinction allows each to function without eclipsing the other.

Like Hawhee, Yameng Liu also focuses on postmodernism's challenge to the discovery/creation binary. In "Invention and Inventiveness," Liu deals more specifically with the institutionalization of modernist values that have secured this opposition, locating the source of invention as discovery in the thought of Francis Bacon and invention as creation in Samuel Taylor Coleridge's romantic conception of discourse production. Liu explores the dependence of both views of invention on a stable, sovereign subject and examines Derrida's deconstruction of the discovery/creation binary. In place of this binary, Liu offers the notion of "inventiveness," which he describes as a "strike for the new without attempting a clear severance with the old." Liu maintains that the deconstruction of the discovery/creation binary holds specific implications for rhetorical pedagogy. It forces us to reexamine the notions of originality and creativity we convey to students and to rethink not only our conceptions of the speaker/writer but also the audience.

In "Institutional Invention: (How) Is It Possible?" Louise Wetherbee Phelps argues that rhetorical invention provides a useful paradigm for understanding constraints on institutional change in academic contexts. She observes that rhetoric plays several roles in institutional development: as a means of change, as a language for explaining invention, and as a feature of its organization as a system. Phelps draws on a number of sources in addition to rhetorical theory: the history of education, theories of institutional and organizational behavior, and her own experiences in educational administration. She maintains that invention by institutions, especially academic ones, can be developed as a practical art, one that not only enables academic institutions to be environments that foster creativity but also allows institutions to reinvent themselves in creative and productive ways.

In "Conflict in Community Collaboration," Linda Flower and Julia Deems bring invention to bear on rhetoric's civic function by examining the use of heuristics in community problem solving. They offer the most explicit discussion of practical heuristics as they explore a series of discussions between tenants and low-income landlords at the Community Literacy Center in Pittsburgh's Northside. Their article records participants' own inventive processes and goes on to examine how heuristics based on the group's deliberative processes helped them to create a discursive space between antagonism and forced consensus. In particular, Flower and Deems discuss the heuristic use of "scenarios," which they describe as "what if" narratives that help participants expand the range of options for community problem solving. As Flower and Deems describe it, invention in this context calls for a generative, openly deliberative, and non-adversarial rhetoric. Such a rhetoric must be capable of not only articulating differences but also engaging in collaborative planning, constructing multivocal meanings, and gaining consensus about actions, if not ideas.

In a different context, Donald Lazere also seeks to restore rhetoric's civic function to the teaching of writing. In "Invention, Critical Thinking, and the Analysis of Political Rhetoric," he argues that effective engagement in public rhetoric requires the ability not only to create but also to analyze public discourse. Lazere maintains that backlash from the academic culture wars has put pressure on writing teachers to take politics out of the classroom, but he insists that participating in rhetoric's public function requires understanding the terms and strategies of contemporary political argument. Lazere contends that engaging students in analyzing political rhetoric fosters critical thinking, as it requires such skills as distinguishing fact from opinion, identifying assumptions, predicting probable consequences, and recognizing different value orientations and ideologies. Building on his work in teaching the political conflicts, he offers specific strategies for interpreting political arguments. Lazere concludes his essay with a discussion of inventional strategies for creating public arguments. Thus Lazere points to a contemporary rhetoric of political discourse, one that is responsible at the same time that it enables dissent and the expression of strong convictions.

Another perspective on rhetoric's civic function is offered by Jay Satterfield and Frederick Antczak in "American Pragmatism and the Public Intellectual: Poetry, Prophecy, and the Process of Invention in Democracy." Both scholars acknowledge that postmodern critiques of foundationalism have so focused on the cultural and institutional constraints on knowledge as to problematize the very notion of invention. In the wake of "collapsed foundations," they offer pragmatism as an ethical and epistemological paradigm that can underwrite a post-foundationalist theory of invention. In this paradigm, they define invention broadly as "the creation of new thought that is workable, but also sharable." More specifically, they describe inventional theory in the pragmatic tradition as politically effective knowledge created in an historically contingent public space. They find the most complete expression of this paradigm in the thought of Cornel West. According to Satterfield and Antczak, West's pragmatism resists foundationalism's naive conceptions of subjectivity, while still allowing for meaningful knowledge and public action. In this paradigm, invention consists in public dialogues directed toward collective action within politically minded institutions.

Haixia Wang uses the work of a Classical Chinese thinker to examine the ways culture shapes conceptions of invention, subjectivity, social hierarchy, and political possibility. In "Inventing Chinese Rhetorical Culture: Zhuang Zi's Teaching," Wang explains that the philosopher viewed invention in an analogical and dynamic relation to context. For Zhuang Zi, invention is characterized by the acts of analyzing, sorting, and assessing—all of which are contingent upon specific circumstances. Understanding this sense of contingency, according to Wang, is key to understanding the probable nature of the Tao. Wang examines, in particular, Zhuang Zi's conception of *ziran,* which she says can be translated as "spontaneity" and "movements of choice," guided not by rules or wild impulses but by a clear vision of things. Wang maintains that for Zhuang Zi this vision could be possessed by everyone. Thus Zhuang Zi's teachings were in tension with the hierarchical character of China's system of imperial examinations. Wang brings these notions to bear not only on Chinese conceptions of invention but also on the Tiananmen Square tragedy.

The last two articles offer historical perspectives on invention. In "Literacy in Athens During the Archaic Period: A Prolegomenon to Rhetorical Invention," Richard Enos describes how the cultural and economic constraints on literacy in Archaic Greece shaped conceptions of invention in the Classical Period. Enos argues that in Archaic Greece writing was viewed as a labor skill—or craft trade—"used by non-experts for facilitating everyday activities," particularly various forms of record keeping. As Enos explains it, these literate record keepers were not aristocrats; indeed, up to the Classical Period, they were generally members of the lowest census class. Enos describes a division of labor between reading and writing that fell along class lines. In this context, the rhetorical curricula of such rhetoricians as Isocrates were revolutionary because they defined writing as a heuristic that facilitated creativity and encouraged intellectual complexity. Enos's discussion is important for providing a more complete context for understanding both the development of rhetoric's public function and the debates between Plato, Isocrates, and Aristotle concerning writing and rhetoric.

"Vico's Triangular Invention," by Mark Williams and Theresa Enos, explores the eighteenth-century rhetorician's theories of knowledge, rhetoric, and invention for insights to inform the teaching of invention. They discuss Vico's conception of invention in the context of his debates with Descartes concerning reason and imagination and pay special attention to the ways in which Vico's conceptions of topical invention, common sense, and memory are contingent on both context and history. Williams and Enos illustrate how Vico's unique integration of these disparate elements undermines the opposition of individual imagination and collective consciousness, providing a basis for rhetoric's public function.

These essays reveal scholars' confrontations with the constraints and possibilities that attend contemporary research in invention. Will they help to create a more protected space for the art? That may still be an open question. Early Greek conceptions of invention depicted the art as a process and act of "making a path." To make a path is to enable new perspectives, new points of contact—even new destinations. Perhaps this is a more productive way of envisioning the art—creating spaces, rather than securing them.

Works Cited

Berlin, James. "Rhetoric and Ideology in the Writing Class." *College English* 50 (1988): 477–94.

———. *Rhetoric, Poetics, and Cultures.* Urbana: NCTE, 1996.

Berthoff, Ann E. "Counterstatement." *College Composition and Communication* 23 (1972): 414–16.

———. "The Problem of Problem Solving." *College Composition and Communication* 22 (1971): 237–42.

Bizzell, Patricia. "Cognition, Convention, and Certainty: What We Need to Know About Writing." *Pre/Text* 3 (1982): 213–43.

Bourdieu, Pierre. *Outline of a Theory of Practice.* Trans. Richard Nice. Cambridge: Cambridge UP, 1977.

———. *The Logic of Practice.* Trans. Richard Nice. Stanford: Sanford UP, 1990.

Derrida, Jacques. *The Truth in Painting.* Trans. Geoff Bennington and Ian McLeod. Chicago: U of Chicago P, 1987.

Emig, Janet. *The Composing Processes of Twelfth Graders.* Urbana: NCTE, 1971.

Fish, Stanley. "Antifoundationalism, Theory Hope, and the Teaching of Composition." *The Current in Criticism.* Ed. Clayton Koelb and Virgil Lokke. West Lafayette: Purdue UP, 1987.

Flower, Linda. "Comment and Response." *College English* 51 (1989): 765–69.

Lauer, Janice M. "Heuristics and Composition." *College Composition and Communication* 21 (1970): 396–404.

———. "Response to Ann E. Berthoff." *College Composition and Communication* 23 (1972): 208–10.

LeFevre, Karen Burke. *Invention as a Social Act.* Carbondale: Southern Illinois UP, 1987.

Winterowd, Ross W., ed. *Contemporary Rhetoric: A Conceptual Background with Readings.* New York: Harcourt Brace Jovanovich, 1975.

———. Introduction. *Contemporary Rhetoric* 1–37.

Young, Richard, and Alton Becker. "Toward a Modern Theory of Rhetoric: A Tagmemic Contribution." Winterowd *Contemporary Rhetoric* 123–43.

Perspectives on Rhetorical Invention

Rhetorical Invention

The Diaspora

JANICE M. LAUER

This essay began as an attempt to identify any new developments in scholarship and teaching in rhetorical invention. As I pursued my investigations, however, I started to wonder whether there *were* any new developments. In the collection *Landmark Essays on Rhetorical Invention,* the latest essay included was published in 1986 and the most recent items in the bibliography were two from 1993. In the October 1995 issue of *College Composition and Communication,* the following statement appears in the "Recent Books" section: "What invention was to discussions of rhetoric/composition in the 1970s—a key concept from classical rhetoric through and around which contemporary research and classroom practices were fundamentally renegotiated—'ethos' has become for the 1990s" (123). Did this mean that concern with invention in composition studies was a matter of the past?

To address this question, I set as a benchmark for the "past" composition studies from the sixties on heuristics to conceptions of invention as a social act in the mid eighties (see Lauer and LeFevre). Over that period, numerous publications focused explicitly on invention, constructing it as complex and multifaceted, encompassing theories and practices of discourse initiation, exploration, argument development, judgment formation, intertextual negotiations, and reader positioning. Invention *theory* included studies of epistemologies and cognitive processes, while invention *pedagogy* included strategies, heuristics, tactics, and plans for discursive inquiry and action. Early invention strategies were often proposed as portable guides for invention across fields, subjects, and genres and as general heuristics in areas outside one's expertise or for ill-defined problems (e.g., the tagmemic heuristic, Burke's pentad, clustering, freewriting, or journaling).

Later composition theorists, following studies in psychology, began to foreground discipline-specific heuristics, while still continuing work on general heuristics (see Nickerson, Perkins, and Smith).

More recently, however, books and articles devoted entirely to contemporary invention have been difficult to find. Instead, rhetorical invention has migrated, entered, settled, and shaped many other areas of theory and practice in rhetoric and composition. Richard McKeon delineated a similar set of moves for invention through the long medieval period: "In application, the art of rhetoric contributed during the period from the fourth to the fourteenth century . . . to the canons of interpreting laws and Scripture, to the dialectical devices of discovery and proof, the establishment of the scholastic method, which was to come into universal use in philosophy and theology, and finally the formulation of scientific inquiry" (211). Like invention in the medieval period, invention today can be found in a diaspora of composition areas rather than in discussions labeled "invention." In this essay, therefore, I will assume the role of cartographer, beginning to map out these sites in order to represent the ubiquity of invention rather than to discuss its treatment in depth in any one site. I am constructing this map as a heuristic, a guide for my own and others' further thinking about rhetorical invention. The sites I will visit illustrate, I believe, that work on invention today is implicit, fragmented, located within many rhetoric and composition places of inquiry.

Further, this examination will reveal that a number of earlier emphases in scholarship on invention have either disappeared or been marginalized: the relationship between invention and the writing process, the heuristic function of invention as a kind of thinking that stimulates new knowledge, invention as an art or strategic practice, the importance of classroom attention to invention, interdisciplinary theoretical linkages with inventional epistemologies, and the consequential nature of invention studies for practice and pedagogy. What this excursion into the diaspora indicates are different dimensions and preoccupations with invention in the last two decades.

Writing in the Disciplines and Writing across the Curriculum

In the first site, Writing in the Disciplines, scholars have been exploring the epistemological role of rhetoric in different disciplines such as biology, music, and management, gradually building an understanding of how these diverse fields make knowledge (e.g., Bazerman; Hoger; Smart). Although I take this body of research to have an inventional cast, these scholarly accounts of epistemological practices have not often trickled down into inventional strategies for the classroom. The emphasis in Writing Across the Curriculum (WAC) courses has been largely on helping students use writing to learn the material in a field rather than to create new knowledge in a discipline. Judith Langer and Lee Odell underscore this point. Langer cites the work of Odell, Charles Bazerman, Patricia Bizzell, and Anne Herrington as those who have revealed the structure of disciplinary thought in different fields, including warrants and claims in particular contexts, but she laments that many teachers in different disciplines focus on content, not on higher-level intellectual skills (71–72). In her study of several classes, she saw teachers beginning to introduce inventional activities in biology, history, and literature. She and her coresearchers found that these teachers were increasingly focusing on the tentative nature of "truth" and emphasizing active questioning and interpretation (72). The researchers noted that biology teachers were preoccupied with challenging the independence of method and observation, while history teachers were debating the most appropriate methods of inquiry (73). She adds an important point, though, when she notes that these teachers' efforts to introduce students to the process of science through their own attempts did not include providing students with any procedural knowledge to apply such methods successfully themselves (75). Odell concurs. Referring to the Writing Across the Curriculum literature, he comments that although "some scholars are beginning to argue that different academic disciplines, even different courses within a discipline, make quite different conceptual demands on writers" (89), several obstacles exist to teaching inventional strategies. He identifies two of these obstacles: teachers may have so internalized their thinking strategies that they find it difficult to make them explicit, or teachers may be more comfortable discussing the content of their disciplines than in identifying methods of thinking or analytic strategies needed to generate or reflect on that content (97).

Another problem facing inventional instruction across the curriculum is introduced by Donna LeCourt, who worries that teaching knowledge-making practices may "serve to reinforce the ways of thinking and status of a particular knowledge" (392). She argues that Writing in the Disciplines studies are not "conventionally innocent; instead they serve as reflections and are constitutive of the ways of knowing and the modes of inquiry valued within certain disciplines" (393). She contends that the goal of WAC efforts is almost always to initiate students into a "certain way of thinking valued by the discipline . . . to 'train' students to think and write within a certain discourse community" (393), allowing "the discourse to appear natural and pragmatic, and thus ideologically free" (395). She advocates instead "positioning the student as an active partner in a dialectic with such communities, making space for the personal—for difference—within the disciplinary" (397). Her pedagogy involves having students conduct investigations into their majors using strategies of cultural critique, asking what the discourse includes and excludes (399). Judy Kirscht, Rhonda Levine, and John Reiff address these issues by discussing the rhetoric of inquiry, which they argue can connect composing to learning and writing in the disciplines. They advocate teaching the rhetoric of inquiry, which entails a growth in dialectical thinking that helps students move "beyond the boundaries of previous belief systems and in exploring new perceptions" within their fields (374). Thus they argue that writing instruction becomes a way not only to "interact with declarative knowledge, but also to develop procedural knowledge concerning the field—to learn *how* knowledge has been constructed as well as *what* that knowledge is" (374).

Public Contexts and Cultural Studies

In another site, the public or civic realm, cultural studies theorists have been framing new heuristics for cultural critique. For example, in response to concerns that cultural studies directs students to deconstruct but not construct, James Berlin has begun to develop heuristics not only to guide critique but also to help students re-envision their cultures. In an essay in *Rhetoric Review* in 1992, he outlined the following heuristic procedures that directed students to:

> —first locate in their experience the points at which they are resisting and negotiating with the cultural codes they encounter (starting strategy),
> —then locate key terms in any text (print, film, television) in these areas and situate these terms within the structure of meaning of which they form a part,
> —then position terms in relation to binary opposites,
> —next place terms within narrative structural forms suggested by the text, and
> —finally, situate these narratives within larger narrative structures that have to do with economic, political, and cultural formations. (26–31)

After using these heuristics to analyze texts, students were directed to apply the guides to their personal experiences and to locate points of conflict or dissonance that warranted similar scrutiny. As Berlin explained: "In enacting the composing process, students are learning that all experience is situated within signifying practices, and that learning to understand personal and social experience involves acts of discourse production and interpretation, the two acting reciprocally in reading and writing codes" (31). The heuristics Berlin introduced were both field specific and yet portable to any subject within that field. Others working in cultural studies have not as explicitly articulated their critical practices as inventional guides, however, nor have they developed their heuristic power to construct new cultural codes to replace those that have been critiqued.

Donald Jones finds problematic the fact that Berlin's students thought they had to "reject their past experiences and ideas as products of the hegemonic culture" (85) and thus grant less credence to their prior beliefs. Jones worries that these instructors offer few specific alternatives to students' present beliefs and thus students are unwilling to submit to what he calls pervasive skepticism. Maintaining that classes do not provide an adequate account of agency and its achievement, he turns to Dewey for a non-foundational pragmatism—an alternate method of inquiry with which students can learn to question, revise, and affirm assertions (86). His pedagogy advocates teachers' posing questions for students to answer about the cultural influences in their personal lives, about "cultural assumptions perpetuated by language" (91), and discursive practices that have coded their experience. Unlike Berlin's, Jones's pedagogy calls for inventional agency and positions teachers, rather than students, as question posers. In any case, both of these compositionists give explicit attention to invention in cultural studies.

Like those in Writing in the Disciplines, composition researchers are also beginning to examine the kinds of thinking entailed in the construction of public discourse in multiple contexts. Some recent dissertations examine the discursive practices of lay and professional writers who compose public discourse. Karen Griggs studies the complex inventional practices of multiple authors (professionals, public representatives, government officers) who collaborated to write the Water Safety Act in Indiana. Karen Dwyer investigates the heuristics used by citizens who write for Amnesty International. Haixia Wang studies the tactics by which the *People's Daily* in China constructed the Tiananmen Square incident for the Chinese public. Thomas Moriarty probes the discursive practices that contributed to a peaceful transition in South Africa. In each of these studies, heuristics play important roles in constructing new understandings in the public sphere. This kind of research offers the field needed knowledge on which to build investigative strategies that can be taught to help students prepare for writing as citizens in different spheres.

Studies of Gender, Race, and Cultural Difference

Another cluster of inventional locations on my map can be found in the vicinity of studies of gender, race, and cultural difference. Gender differences have received considerable attention. Feminist studies of women's ways of knowing and communicating (e.g., Gilligan; Noddings; Tannen) have been visited by scholars in rhetoric and composition (e.g., Flynn; Hollis; Daumer and Runzo; Bridwell-Bowles), who enumerate various women's ways of composing, such as rejecting persuasion as an act of violence in favor of constructing matrices or wombs; choosing personal narrative over argument; deploying language close to the body; composing personal and emotional discourse; playing with language; focusing on the nonlinear, associative, and inchoate as opposed to the local, hierarchical, and argumentative; and featuring concrete particularities instead of abstract generalizations. In this discourse some inventional practices have been elaborated: journaling (Gannett); collaborative planning (Ede and Lunsford; Flower); dialoguing and interviewing for ideas (Hays); reading women's narratives to illustrate naming oneself instead of being defined by others (Daumer and Runzo; Bridwell-Bowles); and using the believing game as connected learning (Hays). Such efforts to identify features of feminine epistemologies, however, have yet to be adequately coordinated or developed into guides for helping women to develop as writers. Further, these discussions of women's ways of knowing and composing have come up against several critiques or cautions in feminist studies (e.g., Fuss; de Lauretis) and in composition (e.g., Ritchie; Kirsch) that bear on invention. These critiques include (1) the charge of an essentialism, including the tendency to obscure differences in race, class, sexual preference, and ethnicity among women; (2) some disagreement over avoiding conflict and argument (Jarrett); (3) cautions about uncritical uses of collaboration (Ashton-Jones); and (4) concern that women may remain in dualistic or other less complex forms of reasoning (Hays). Although much of this work has focused on analyses of texts rather than on composing processes which admit of arts of inquiry, these studies constitute a fruitful site for further work on invention.

Several areas of scholarship on racial diversity are also resonant with inventional concerns: investigations of oral and literate cognitive styles (e.g., Heath; Scollon and Scollon); accounts of how literacy is achieved (e.g., Moss); personal narratives about literate ways of knowing (e.g., Villaneuva; Rose); and studies of cultural ways of composing or knowing of Native Americans, African Americans, and Hispanics. For example, Jeanne Smith speaks of Lakota students as using highly developed narrative thinking to compose, incorporating metaphors and cultural materials such as songs and legends. Discussing Hispanic/Latino writing, Victor Villaneuva cites a number of research studies that identify features of Hispanic writing such as conscious digressions in logic and use of amplification, what he calls fundamentally sophistic. Jacqueline Jones Royster's study of nineteenth- and early-twentieth-century black women points to "patterns emerging that highlight the tradition of black feminist thought: a penchant for contextualizing and centering pieces of the puzzle of black people's lives in the United States that did not fit and their capacity to image double things within adverse circumstances" (108). Henry Louis Gates extensively analyzes signifyin[g] as the master black trope, subsuming multiple subtypes such as talking smart, putting down, playing the dozens, shagging, and rapping. Gates points out that signifyin[g] is a term for conscious rhetorical strategies, structures at work in the discourse, rather than specific content. Such definitions of signifyin[g] position it as a complexus of inventional arts that guides the discourser to construct meanings and to accomplish powerful ends skillfully in different contexts. Gates further explains that signifyin[g] is a form of rhetorical training, whose most sublime uses are taught by an indirect mode of narration, by adolescents manipulating these black figures, learning to move freely between two discursive universes. Mastering these signifyin[g] practices creates, in Gates' terms, *homo rhetoricus Africanus*. Further, he identifies the Monkey, the source and encoded keeper of Signifying, as the figure of *writing*.

The above studies have identified and analyzed epistemological and rhetorical differences, but only a few have addressed how inventional powers are honed or have outlined practices and pedagogies to guide African American, Hispanic, or Native American students in further developing their ways of knowing and inventing.

In current cross-cultural studies, some interest in invention can also be detected. Ulla Connor reviews research on the texts of Arabic, Chinese, Japanese, Korean, Finnish, Spanish, and Czech writers, primarily describing differences in the finished products such as syntactic patterns, placement of thesis, organizational structure, types of discourse, and formality and elaboration of language—differences that one might infer have been generated by inventional processes. She also reports the results of studies on the use of the classical appeals and Toulmin's argument model in the compositions of international students. The IEA study of written discourse in seventeen countries edited by T. P. Gorman, A. C. Purves, and R. E. Degenhart has a similar textual emphasis, but its general model of discourse does include kinds of cognitive processing such as inventing/generating, organizing/reorganizing, and reproducing.

In these cross-cultural studies as well as those on gender and race, important distinctive features have been identified, but this work has not moved very far into the realm of strategic knowledge to develop guides or arts that facilitate these epistemologies. Thus the consequential nature of this scholarship has yet to be fully exploited for pedagogy and practice.

Theories of Technology

Interest in invention has also migrated to the site of theory and research on technology. Early work (e.g., Burns and Culp) largely positioned existing inventional heuristics on-line, enabling some interaction between the computer and the students. More recent work has begun to examine unique ways of constructing knowledge through technology and includes studies on the role of the visual in the making of meaning and the interface between the verbal and the visual in knowledge construction facilitated by computers (Sullivan). Barbara Mirel makes a case for the value of tabular reports, enabled through technology, in transforming raw data into meaningful information, an inventional process entailing a dynamic interplay between writer's rhetorical and technical skills. Ron Fortune explores the role of the computer in finding ways to stimulate visual and verbal thought processes. Johndan Johnson-Eilola and others are discussing the inventional powers of hypertext, helping writers to develop their own heuristics. In *Opening Spaces: Writing Technologies and Critical Research Practices,* Patricia Sullivan and James Porter argue for the use of postmodern mapping as a methodological heuristic. While few of these works use the term "invention," the import and potential are evident.

Studies of Genre

Genre studies are an ironic place to look for invention for those of us who introduced conceptions of invention and the composing process in the sixties. One obstacle we faced was a fixation in composition instruction with the modes, the reigning genre configuration. The modes had domesticated invention, indenturing it to serve EDNA (Exposition, Description, Narration, Argumentation), to foster discrete methods of development, a fate that Sharon Crowley has documented. Not a heuristic but methodical memory, invention was assigned to convey the known. To extricate invention from this managerial rhetoric, heuristics were developed in the sixties and seventies to help writers go beyond the known, to guide writing to create new understanding rather than to fill in the blanks of a mode.

Recent genre theories, however, offer invention more suitable homes. Let me turn here to three examples from the work of Carolyn Miller, Mikhail Bakhtin, and Carol Berkenkotter and Thomas Huckin. In what sense does their work suggest, if not fully articulate, new conceptions of invention? When arguing for a rhetorical basis for classifications of discourse, Miller posits genres as rhetorically sound if they reflect "the rhetorical experience of the people who create and interpret discourse" (152), that is, if they encompass the inventional processes of discourse construction. She argues for a view of genre as action (symbolic or otherwise), situated, attributing motives (155–56), a conception that seeks to explicate the knowledge that practice creates. One of her bases for discriminating among genres is their exigencies, which she defines as recurrent intersubjective phenomena in contrast to Bitzer's sense of rhetorical situation as outside the rhetor. Thus I consider her to be classifying genres on an inventional basis because historically one of the key inventional acts has been the initiation of discourse through a construction of exigency. Miller enriches our conception of exigency, calling it social knowledge, "a mutual construing of objects, events, interests, and purposes that not only links them but makes them what they are: an objectified social need" (157), embodying an aspect of cultural rationality (165). She theorizes that compelling situations are social constructs, not the result of individual perception, but of intersubjective definition and classification.

Bakhtin's theory of genre also makes room for invention. For example, in his discussion of the finalization of the utterance (what he calls the inner side of the change of speech subjects), one means of guaranteeing the possibility of response is the semantic exhaustiveness of the subject, which he maintains varies greatly in different spheres of life. In explaining how one reaches this exhaustiveness in the action of creation, he advances the notion of the speech plan or will that determines the boundaries, length, and choice of subject and generic form (76), a plan that sounds quite inventional to me.

Finally, Berkenkotter and Huckin argue that writers acquire and strategically deploy genre knowledge as they participate in their field's or profession's knowledge-making activities (477). They conceive of genre knowledge as kairotic, including a sense of what is appropriate to a particular purpose in a particular situation or time (478). Further, Berkenkotter and Huckin maintain that genre knowledge is inextricable from professional writers' procedural and social knowledge (487), as well as their knowledge of appropriate topics and relevant details (488).

In all three of these theoretical conceptions of genre, I find inventional leanings because genre is viewed as social action, entailing meaning construction. But despite these reconceptions of genre, invention is still EDNA's servant in many classrooms and textbooks, acting only to find material to develop types of discourse.

Hermeneutics

One of the long-standing conversations in composition studies has been the relationship between the creative and interpretive acts, between heuristics and hermeneutics. I will enter this conversation by responding to an essay in *Rhetorical Hermeneutics* by Dilip Gaonkar, who contrasts classical rhetoric and contemporary rhetoric on this point. Classical theories, he contends, gave priority to rhetoric as a practical/productive/cultural activity in the civic realm, viewing the "rhetor as (ideally) the conscious deliberating agent who chooses and discloses the capacity for prudence, who invents discourse that displays an ingenium, reducing the agency of the rhetoric to the conscious and strategic thinking of the rhetor" (26–49). Contemporary rhetoric, on the other hand, he argues, "extends the range of rhetoric to include discourse types such as scientific texts and gives priority to rhetoric as a critical/interpretive theory" (26). As a consequence of this move, he maintains that "First, what is rhetorical in any given case is invariably an effect of one's reading rather than a quality intrinsic to the object being read. . . . Hence the range of rhetoric is potentially universal" (29). Concluding that contemporary rhetoric has moved from a vocabulary of production to a vocabulary of reception, he wonders: "Is it possible to translate effectively an Aristotelian vocabulary initially generated in the course of "theorizing" about certain types of practical (praxis) and productive (poesis) activities delimited to the realm of appearances (that is, "public sphere" as the Greeks understood it) into a vocabulary for interpretive understanding of cultural practices that cover the whole of human affairs, including science?" (30).

The binary Gaonkar constructs separates invention from hermeneutics. But this binary doesn't fit our experience in composition, which Gaonkar ignores. Since the sixties, composition studies of invention have encompassed both the classical and contemporary senses of rhetoric, because composition has been in the business of theorizing and teaching a productive art, helping students to construct meaning in discourse instead of only transmitting meaning through style. In composition studies over the years, inventional theory has been socialized, complicated, and distributed, but it has not been exiled as a productive art.

Since the sixties, collections of essays or "readers" have remained a staple of composition classes while hermeneutic theory has undergone several transformations. As Berlin put it in his last book: "The teacher's most demanding, engaging, and creative acts, then, are the encouraging of complex reading and writing strategies and practices. Students must learn the signifying practices of text production . . . as well as the signifying practices of text reception" (111). While the relative merits of reading and writing strategies as inventional arts continue to be debated, heuristics and hermeneutics have remained linked in classrooms, an experience of practice and pedagogy that Gaonkar doesn't take into account.

In conclusion, the above map of the diaspora I have just constructed sketches recent inventional studies and pedagogies as dispersed and localized, precluding any final characterization of a unified theory or common sets of practices. While most sites are paying attention to diverse ways of knowing, discussions are primarily aimed at explaining the *nature* of different epistemologies (the kind of knowledge that Aristotle would have called *episteme* in contrast to productive knowledge or *techne*). These two kinds of knowledge interact, however, because developing arts of invention entails understanding the nature of the heuristic acts they will facilitate. This kind of interaction has long flourished in rhetoric and composition studies, which encourages both kinds of intellectual work. With these new understandings of diverse epistemologies, the field is now in a position to offer students discipline-specific heuristics and situated inquiry practices for the workplace and public realm. We can now develop guides for argument, investigation, and critique that are sensitive to cultural, gender, and racial differences. We can further integrate invention with technology and more deeply explore the relationship between heuristics and hermeneutics. The diaspora offers a promising terrain for future construction of multiple and rich arts of invention.

Works Cited

Ashton-Jones, Evelyn. "Conversation, Collaboration, and the Politics of Gender." *Feminine Principles and Women's Experience in American Composition and Rhetoric*. Ed. Janet Emig and Louise Phelps. Pittsburgh: U of Pittsburgh P, 1995. 5–26.

Bakhtin, M. M. *Speech Genres and Other Late Essays.* Trans. Vern McGee. Ed. Caryl Emerson and Michael Holquist. Austin: U of Texas P, 1986.

Bazerman, Charles. *Shaping Written Knowledge: The Genre and Activity of the Experimental Article in Science.* Madison: U of Wisconsin P, 1982.

Berkenkotter, Carol, and Thomas Huckin. "Rethinking Genre from a Sociocognitive Perspective." *Written Communication* 10 (1993): 475–509.

Berlin, James. "Poststructuralism, Cultural Studies, and the Composition Classroom: Postmodern Theory in Practice." *Rhetoric Review* 11 (1992): 16–33.

———. *Rhetoric, Poetics, and Cultures: Refiguring College English Studies.* Urbana: NCTE, 1996.

Bridwell-Bowles, Lillian. "Discourse and Diversity: Experimental Writing Within the Academy." *Feminine Principles and Women's Experience in American Composition and Rhetoric*. Ed. Janet Emig and Louise Phelps. Pittsburgh: U of Pittsburgh P, 1995. 43–66.

Burns, Hugh, and G. Culp. "Stimulating Invention in English Composition through Computer-Assisted Instruction." *Educational Technology* 20 (1980): 5–10.

College Composition and Communication 46 (1995): 456.

Connor, Ulla. *Contrastive Rhetoric: Cross-Cultural Aspects of Second-Language Writing.* New York: Cambridge UP, 1996.

Crowley, Sharon. *Methodical Memory: Invention in Current-Traditional Rhetoric.* Carbondale: Southern Illinois UP, 1990.

Daumer, Elizabeth, and Sandra Runzo. "Transforming the Composition Classroom." *Teaching Writing.* Ed. Cynthia Caywood and Gillian Overing. Albany: State U of New York P, 1986. 45–62.

de Lauretis, Theresa. "The Essence of the Triangle or, Taking the Risk of Essentialism Seriously: Feminist Theory in Italy, the U.S., and Britain." *Differences: A Journal of Feminist Cultural Studies* 1 (1989): 3–37.

Dwyer, Karen. *A Cultural and Rhetorical Analysis of Internationalized Human Rights Discourse.* Diss. Purdue U, 1997.

Ede, Lisa, and Andrea Lunsford. *Singular Texts, Plural Authors: Perspectives on Collaborative Writing.* Carbondale: Southern Illinois UP, 1990.

Flower, Linda. "Strategic Knowledge and the Logic of the Learner." *The Construction of Negotiated Meaning: A Social Cognitive Theory of Writing.* Carbondale: Southern Illinois UP, 1994. 192–222.

Flynn, Elizabeth. "Composing as a Women." *College Composition and Communication* 39 (1988): 423–35.

Fortune, Ron. "Visual and Verbal Thinking: Drawing and Word-Processing Software in Writing Instruction." *Critical Perspectives.* Ed. Gail Hawisher and Cindy Selfe. New York: Teachers College P, 1989. 145–86.

Fuss, Diana. *Essentially Speaking: Feminism, Nature, and Difference.* New York: Routledge, 1989.

Gannett, Cinthia. *Gender and the Journal: Diaries and Academic Discourse.* Albany: State U of New York P, 1992. 19–42.

Gaonkar, Dilip. "The Idea of Rhetoric in the Rhetoric of Science." *Rhetorical Hermeneutics: Invention and Interpretation in the Age of Science.* Ed. Alan Gross and William Keith. Albany: State U of New York P, 1997. 60–102.

Gates, Henry Louis. "The Signifying Monkey and the Language of Signifyin[g]: Rhetorical Difference and the Orders of Meaning." *Signifying Monkey: A Theory of Afro-American Literary Criticism.* New York: Oxford UP, 1988.

Gilligan, Carol. *In a Different Voice: Psychological Theory and Women's Development.* Cambridge: Harvard UP, 1982.

Gorman, T. P., A. C. Purves, and R. E. Degenhart, eds. *The IEA Study of Written Composition I: The International Writing Tasks and Scoring Scales.* New York: Pergamon, 1988.

Griggs, Karen. "Audience Complexities in Administrative Law: An Historical Case Study of an Environmental Policy." Diss. Purdue U, 1994.

Hays, Janice. "Intellectual Parenting and a Developmental Feminist Writing Pedagogy." *Feminist Principles and Women's Experience in American Composition and Rhetoric.* Ed. Janet Emig and Louise Wetherbee Phelps. Pittsburgh: U of Pittsburgh P, 1995. 153–90.

Heath, Shirley. *Ways with Words: Language, Life and Works in Communities and Classrooms.* New York: Cambridge UP, 1983.

Hoger, Elizabeth. *Writing in the Discipline of Music: Rhetorical Parameters in Writings about Music Criticism.* Diss. Purdue U, 1992.

Hollis, Karyn. "Feminism in Writing Workshops: A New Pedagogy." *College Composition and Communication* 43 (1992): 340–48.

Jarrett, Susan. "Feminism and Composition: The Case for Conflict." *Contending with Words.* New York: MLA, 1991. 105–23.

Johnson-Eilola, Johndan. *Nostalgic Angels: Rearticulating Hypertext Writing.* Norwood: Ablex, 1996.

Jones, Donald C. "Beyond the Modern Impasse of Agency: The Resounding Relevance of John Dewey's Tacit Tradition." *Journal of Advanced Composition* 16 (1996): 81–102.

Kirsch, Gesa. *Women Writing in the Academy: Audience, Authority, and Transformation.* Carbondale: Southern Illinois UP, 1993.

Kirscht, Judy, Rhonda Levine, and John Reiff. "Evolving Paradigms: WAC and the Rhetoric of Inquiry." *College Composition and Communication* 45 (1994): 369–80.

Langer, Judith. "Speaking of Knowledge: Conceptions of Understanding in Academic Disciplines." *Writing, Teaching, and Learning in the Disciplines.* Ed. Anne Herrington and Charles Moran. New York: MLA, 1992. 69–85.

Lauer, Janice. "Heuristics and Composition." *College Composition and Communication* 21 (1970): 396–404.

LeCourt, Donna. "WAC as Critical Pedagogy: The Third Stage?" *Journal of Advanced Composition* 16 (1996): 389–405.

LeFevre, Karen. *Invention as a Social Act.* Carbondale: Southern Illinois UP, 1986.

McKeon, Richard. "Rhetoric in the Middle Ages." *The Province of Rhetoric.* Ed. Joseph Schwartz and John Rycenga. New York: Ronald, 1965. 172–212.

Miller, Carolyn. "Genre as Social Action." *Quarterly Journal of Speech* 70 (1984): 151–62.

Mirel, Barbara. "Writing and Database Technology: How Writing Helps Extend the Definition of the Workplace." *Electronic Literacies in the Workplace: Technologies of Writing*. Ed. Jennie Dautermann and Patricia Sullivan. Urbana: NCTE, 1996.

Moriarty, Thomas. "South Africa's Rhetoric of Reconciliation: Changes in ANC and Pretoria Government Rhetoric, 1985–1991." Diss. Purdue U, 1998.

Moss, Beverly, ed. *Literacy Across Communities.* Cresskill: Hampton, 1994.

Nickerson, Raymond, David Perkins, and Edward Smith. *The Teaching of Thinking.* Hillsdale: Erlbaum, 1985.

Noddings, Nell. *Caring: A Feminine Approach to Ethics and Moral Education.* Berkeley: U of California P, 1984.

Odell, Lee. "Context-specific Ways of Knowing and the Evaluation of Writing." *Writing, Teaching, and Learning in the Disciplines.* Ed. Anne Herrington and Charles Moran. New York: MLA, 1992. 86–98.

Ritchie, Joy. "Confronting the 'Essential' Problem: Reconnecting Feminist Theory and Pedagogy." *Journal of Advanced Composition* 10 (1990): 249–73.

Rose, Mike. *Lives On the Boundary: The Struggles and Achievements of America's Under Prepared.* New York: Free Press, 1989.

Royster, Jackie Jones. "Perspectives on the Intellectual Tradition of Black Women Writers." *The Right to Literacy.* Ed. Andrea Lunsford, Helene Moglen, and James Slevin. New York: MLA, 1990. 103–12.

Scollon, R., and S. Scollon. *Narrative, Literacy and Race in Interethnic Communication.* Norwood: Ablex, 1981.

Smart, Graham. "Writing to Discover and Structure Meaning in the World of Business." *Carleton Papers in Applied Language Studies* 2 (1985): 33–44.

Smith, Jeanne. "Native American Composition." *Encyclopedia of English Studies and Language Arts.* General Ed. Alan Purves. New York: Scholastic, 1994.

Sullivan, Patricia. "Visual Markers for Navigation Instructional Texts." *Journal of Technical Writing and Communication* 20 (1990): 255–67.

Sullivan, Patricia, and James E. Porter. *Opening Spaces: Writing Technologies and Critical Research Practices.* Greenwich: Ablex, 1997.

Tannen, Deborah. *You Just Don't Understand: Women and Men in Conversation.* New York: Oxford UP, 1990.

Villanueva, Victor. "Hispanic/Latino Writing." *Encyclopedia of English Studies and Language Arts.* General Ed. Alan Purves. New York: Scholastic, 1994.

Wang, Haixia. "Chinese Public Discourse: A Rhetorical Analysis of Account of the Tiananmen Square Incident by the Newspaper 'People's Daily.'" Diss. Purdue U, 1993.

Young, Richard, and Yameng Liu, eds. *Landmark Essays on Rhetorical Invention.* Davis: Hermagoras, 1994.

Kairotic Encounters

DEBRA HAWHEE

In their introduction to *Landmark Essays on Rhetorical Invention,* Richard E. Young and Yameng Liu outline two conflicting perspectives in early scholarship on invention. The first views invention as a process of discovery and posits a belief "in a preexistent, *objective* determining rhetorical order whose grasp by the rhetor holds the key to the success of any symbolic transaction" (xiii). The second perspective views rhetorical invention as a creative process, emphasizing "a generative *subjectivity*" in which discursive production depends on the rhetor's ability to produce arguments. The distinction between these two perspectives hinges on issues of exteriority and interiority: the discovery model presents a subject that looks outside itself to "find" arguments, and the creative model assumes that the subject need only look inside itself for things to say. Despite divergent epistemologies, however, both perspectives depend on a particular model of subjectivity. That is, both the "grasping" and "generating" models of invention posit an active, sovereign subject who sets out either to "find" or "create" discursive "stuff."

For the last twenty years or so, postmodern perspectives have problematized such assumptions about subjectivity and discourse. As Lester Faigley points out, scholars in composition and rhetoric have begun to interrogate the notion of a discrete, sovereign rhetor. Still, as Faigley contends, "where composition studies has proven least receptive to postmodern theory is in surrendering its belief in the writer as an autonomous self" (15). Perhaps this is so because problematizing subjectivity forces a reconceptualization of rhetoric itself, disrupting the rhetorical triangle and calling into question such commonplaces as "rhetoric is persuasion." The issue of postmodern subjectivity and rhetorical pedagogy is so broad and far-reaching that it cannot be fully dealt with in a book, let alone an article. Nevertheless, this issue will inform my investigation of invention. Given the incommensurability of current models of invention and emerging views of subjectivity and discourse, how might alternative models of subjectivity reconfigure rhetorical invention? What effects would such a reconfiguration have on rhetoric?

Invention-in-the-Middle

The concept of invention comes from the Greek verb *heuriskô*, which may be translated "I discover" or "I find." In Greek literature, the verb most often occurs in its active forms, so that an agent or subject discovers or finds a particular object. Yet occasionally the verb takes on what philologists refer to as the middle voice, a reflexive grammatical construction which conflates active and passive meanings: the subject at once also becomes the object. Sophocles' *Electra* presents a rather interesting use of the middle form: *erga tous logous heurisketai,* or, "and deeds make themselves words" (625; trans. L&S). *Electra*'s use of the middle form of *heuriskô* shifts the emphasis of the word away from its traditional unidirectional, object-targeted meaning (I find or discover x) to more bidirectional, reflexive movement: deeds make themselves words. I would like to consider, for a moment, the possibility of what I term "invention-in-the-middle" as an alternative to the distinctly objective and subjective models outlined by Young and Liu.

As a grammatical construct in ancient languages, the middle voice literally falls between the active and the passive. As Eric Charles White puts it, "the 'middle voice' can be understood as a possibility of movement among active, passive, and reflexive forms" (52).[1] A consideration of invention-in-the-middle presses on the issue of subjectivity from a slightly different angle than approaches that suggest rhetors either discover or create. The discovery and creation models both depend on active constructions that presuppose a subject that is better described as the *outcome* of the rhetorical situation. In other words, when "discovering" or "making" arguments, one also "makes" a rhetorical subject. That is, as a first move, the discursive encounter itself forges a different subject; and as a second move, the emergent subject becomes a force in the emerging discourse: "I invent" in the middle becomes "I invent and am invented by myself and others" (in each encounter). The middle, then, at once combines and exceeds the forces of active and passive. In the middle, one invents and is invented, one writes and is written, constitutes and is constituted.

The sophist Gorgias exemplified "invention-in-the-middle" when he took the stage at the Athenian theater and challenged the audience to "suggest a subject," a move which, according to Diogenes Laertius, showed that he "would trust to the moment (*tôi kairôi*) to speak on any subject" (DK 82A1a; Sprague 31).[2] In other words, Gorgias would exploit the possibilities immanent in a particular rhetorical moment—a *kairos*—to create a discursive offshoot, and along with that a new ethos to go somewhere else. "Invention-in-the-middle" occurs always on the spur of the moment, as a response to the forces at work in a particular encounter. The sophistic concept *kairos* thus becomes critical to discourse production, as it marks the opportunity for a subject to produce discourse, even as it marks the other side of subjection—i.e., one is *called upon* to produce discourse.

As the ancient conception of time that attends to degrees of propitiousness, *kairos* does not have a direct English equivalent. Most frequently translated as "exact or critical time, season, opportunity" (L&S 859), *kairos* marks the quality of time rather than time's quantity, which is captured by the other, more familiar Greek word for time, *chronos*. In short, *chronos* marks duration while *kairos* marks force. *Kairos* is thus rhetoric's time, for the quality, direction, and movement of discursive encounters depend more on the forces at work on and in a particular moment than their quantifiable length.[3]

Elaborating *kairos* as a major principle of sophistical rhetoric, John Poulakos writes that the concept realizes "that speech exists in time and is uttered both as a spontaneous formulation of and a barely constituted response to a new situation unfolding in the immediate present" (61). *Kairos,* then, enables a consideration of "invention-in-the-middle," a space-time which marks the emergence of a pro-visional "subject," one that works *on*—and is worked on *by*—the situation. The sophistical notion of *kairos* offers a tool through which to articulate invention away from notions of rhetorical *beginnings,* with specific "ends" in sight (persuasion, for example), and toward notions of discursive *movements,* the in-betweenness of rhetoric. In other words, to paraphrase White, *kairos* necessitates that thought always be on the move to resist freezing (41).

This essay seeks to sketch the contours of an "invention-in-the- middle" by trafficking between the concept of *kairos* and the discourse that congeals around the figure of Gorgias. What emerges herein is not a complete "theory" of invention, but rather more of an intervention into the discourse on invention, a retooling of rhetorical invention. I will elaborate "invention-in-the-middle" by laying out the assumptions about discourse and rhetoric on which this mode of invention depends: sophistic movement, the realm of the between, and *logos* as *dunamis.*

Sophistic Movement

Invention-in-the-middle is a kairotic movement, a simultaneous extending outward and folding back. This space between subject and object is mediated by action—in *Electra*'s case, a making of deeds into words. Consider the mythical figure Kairos, who was depicted as a well-muscled wing-footed figure perched on a stick or ball, balancing a set of scales on a razor blade (fig. 1). Himerius claims that the fourth-century B.C.E. sculptor Lysippus "enrolled Kairos among the gods" (qtd. in Cook 860). According to Pausinias, Ion of Chios called Kairos the youngest son of

Fig. 1. The mythical figure Kairos was depicted as a well-muscled wing-footed figure perched on a stick or ball, balancing a set of scales on a razor blade. His long forelock suggests the importance of looking out for kairos, as the propitious moment, and grabbing it before he passes by.

Zeus (859), making him the younger brother of Athena, the wily goddess of military encounters and of the practical arts.

The god Kairos is most often depicted (as in fig. 1) with winged feet, sometimes with wings on his back as well. He sports an athletic build, of the type Jean-Pierre Vernant calls the "divine super body" (23), with visible back and shoulder muscles and bulging calves. Kairos is often depicted as mostly bald with a long forelock; of all his attributes, his coiffure garners the most attention. Geoffrey Whitney, in his 1586 portrayal of Lyssipus' Kairos (now latinized and bearing the feminine name Occasio or Occasion), explains the significance of the unusual hairstyle (the voice of Occasion is italicized):

> What meanes longe lockes before? *that suche as meete,*
> *Maye houlde at firste, when they occasion finde.*
> Thy head behinde all balde, what telles it more?
> *That none shoulde houlde, that let me slippe before.* (181)

This depiction of the god Kairos (fig. 1) suggests the importance of looking out for *kairos,* as the propitious moment, and grabbing his forelock before he passes by to avoid being left to swipe at a bald head. Offering similar emblematic qualities is the portrayal of Kairos on a Theban limestone relief (now at Cairo), which features a winged figure dressed in military garb running above two female figures: one is dejectedly sitting in the background, while the other is flying ahead in the foreground (fig. 2). J. Strzygowski calls them Kairos, Pronoia, and Metanoia (Cook 863). Pronoia, the figure of foresight, is flying with Kairos, while Metanoia, the figure of afterthought or hindsight, is left behind. The relief, like Whitney's depiction of Occasion, underscores the fleeting quality of Kairos and the consequences of not being prepared for contingencies or opportunities. This reading is also supported by the early addition of a butterfly—also an emblem of mutability—to Kairos' other hand in the Hellenistic period (860).

These characteristics—hair, butterfly, and wings—support the standard symbolic reading of *kairos* as an embodiment of *carpe diem,* now a timeless literary trope—with a decidedly individualistic motif. Now, however, it seems productive to disengage *kairos* from this symbolic stability and examine the ways in which Kairos' "corporeal codes" enable both new conceptualizations of the relations between humans, gods, and time and new perspectives on the concept of discourse itself.[4] What does the god's corporeal code suggest about discourse? Kairos' winged feet, when considered together with discourse—from the Latin *discurrere,* to run to

Fig. 2. Another telling portrayal of Kairos features a winged figure dressed in military garb running above two female figures. Pronoia, the figure of foresight, is flying with Kairos, while Metanoia, the figure of afterthought or hindsight, is left behind.

and fro, to traverse—foreground movement, the movement of time, the movement of language, and the shifting of forces. *Kairos* thus might be read to mark the celerity and multi-directionality of discourse; as such, the mythical figure bears witness to rhetorical "movement."

The available remnants of discourse attributed to Gorgias and his contemporaries suggest that Gorgias cultivated a kind of sophistic movement —in the sense of mode and tempo rather than historical period. Gorgias emphasized the rhythmic movement of discourse, a movement that was closely tied to the "mobility" of discourse. Philosophers Gilles Deleuze and Félix Guattari, writing about the movements of music, observe that in the musical milieu, "rhythm is critical; it ties together critical moments, or ties itself together in passing from one milieu to another" (313). Gorgias is known for emphasizing the rhythmic movements of discourse. One of his legacies, as it is widely thought, is the introduction of rhythm and poetic style to the art of words. As Diodorus Siculus points out, "[Gorgias] was the first to use extravagant figures of speech marked by deliberate art: antithesis and clauses of exactly or approximately equal length and rhythm and others" (DK 82A4; Sprague 33). Similarly, Suidas contends that Gorgias "was the first to give the rhetorical genre the verbal power and art of deliberate culture and employed tropes and metaphors and figurative language and hypallage and catachresis and hyperbaton and doublings of words and repetitions and apostrophes and clauses of equal length" (DK 82A2; Sprague 32).

Gorgias thus moved lyrical poetry into the realm of rhetoric. As Charles P. Segal notes, "Gorgias, in fact, transfers the emotive devices and effects of poetry to his own prose, and in doing so he brings within the competence of the rhetor the power to move the psyche by those suprarational forces which Damon is said to have discerned in the rhythm and harmony of the formal structures of music" (127). Segal suggests that Gorgias followed the work of his contemporary Damon (n. 103), who studied music's effects on the movement (*kinesis*) of the psyche. Further, Bromley Smith compares Gorgias' speech to "a symphony because when read aloud it recalls a piece of music; for it has the cadences, tonal effects, diminuendos and crescendos of a sonata" (350), and Edward Schiappa argues that a proper title for Gorgias is "prose rhapsode," thus marking Gorgias' "striking and almost musical" style (251, 245) and his hybridized (poetic-prosaic) discursive strategies. Gorgias is thus the most musical of the sophists. Attuned to the effects of rhythms in speech, he becomes a part of the discursive forces at work, as the harmonies of poetry meld with the art of speaking, producing an art (paradoxically) grounded in movement.

Most of the figures Gorgias is credited with having brought to the domain of rhetoric suggest some sort of movement. For example, tropes (*tropais*), from *tropê*, meaning turn, turning, and *tropos,* can be used to indicate musical harmony, or a particular mode. Metaphor, from *metaphora,* transport, haulage, change, even a passing phase of the moon, was itself "transferred" to indicate the "transference of a word to a new sense." Hypallage (*hypallagê*), a term for interchange or exchange, such as the exchange of women, or the change of regime or the color of wine (L&S 1851), came also to denote a verbal play on shifts in shades of meaning; apostrophe from *apostrephein*, a turning away, a bend in the stream, or, in rhetoric, a turning away from others to address one (L&S 220). Most of the figures are verbs-cum-nouns, and most mark the twists and turns (and potential twists and turns) that discourse can produce. The figures Gorgias used as tools of flight to produce discourse are tools of movement, facilitating the kind of discursive action that was later dubbed "to Gorgianize" (Philostratus, qtd. in DK A35; Sprague 41).

Gorgias is thus often characterized as the rhetor of style. Still, Gorgianic "style" is not the ornamental dimension of rhetoric (style as opposed to content). Instead, style, for Gorgias, is the rhythm and movement of the discourse itself; as such, it is irreducibly linked to content (and hence to what is traditionally figured as rhetorical invention). Thus before the canons were neatly divided into five discrete steps for discourse production, Gorgias demonstrated the impossibility of maintaining their separation. In other words, style, for Gorgias, was a means of invention, a kind of movement that Gorgias was always already caught up in, a movement that provides constraints along with possibilities.

In his defense of Helen, a speech in which he interrogates the prevailing assumptions about Helen's responsibility for the Trojan War, Gorgias suggests that the power (*dunamis*) of speech could be the reason for Helen's flight to Troy. In doing so, however, Gorgias also performs his point, implicating himself and his listeners in his own speech. He begins this line of argument by calling speech a "powerful lord" (*dunastês megas*) that can effectively "banish fear and remove grief and instill pleasure and enhance pity" (8).[5] Before he develops this point, Gorgias addresses his hearers in the imperative: "listen (*phere*) as I turn (*metastô*) from one argument (*logon*) to another" (9). The verb *metastô*, from *methistêmi,* here translated as "turn," is a verb of movement. It generally takes the force of "to transform" or "change," as in to change form or position. This moment of direct address thus marks a critical—and literal—turning point in the *Helen:* not only does it mark a transition from one argument to the next, but it marks the transformation of Gorgias himself in that discursive movement. Gorgias does more than catalogue arguments; he cultivates an ethos that morphs between *logoi*. It is, therefore, the *turn* itself, not the *logoi,* but the very act of changing and being changed that Gorgias foregrounds when he directs those present to listen (*phere*). The verb *pherô* means "to bear" or "to carry," but it can also refer, at the same time, to a yielding or producing, as a cow producing (and hence bearing) milk. The act of listening, then, becomes just that: a productive, transformative act for hearers and speakers. At this point, Gorgias orders his listeners both to bear and to produce his act of turning. This moment of direct address, then, emphasizes the transformative encounter produced through discourse.

Gorgias' speech thus inscribes him as a shapeshifter—for the sake of "argument"—performing a kairotic "invention-in-the-middle," by which Gorgias inserts himself into the situation at that particular moment, imploring those present to *phere,* to bear and produce the transformative rhetorical encounter: "Listen as I turn."

Intermezzo

The only way to get outside the dualisms is to be-between, to pass between the intermezzo. . . .

Deleuze and Guattari (1987)

Given its relationship to the movements of discourse, "invention-in-the-middle" resonates with a more general "betweenness," an action that happens in the thick of things. "Invention-in-the-middle" assumes that rhetoric is a performance, a discursive-material-bodily-temporal encounter, a force among forces. This mode of invention is not a beginning, as the first canon is often articulated, but a middle, an in-between, a simultaneously interruptive and connective hooking-in to circulating discourses.

As an interruption, "invention-in-the-middle" calls for rhetorical cuttings, interventional piercings of particular moments to produce discourse. It is important to distinguish "invention-in-the-middle" and its partner term *kairos* from what is often called exigence. While a commonly held notion of exigence requires the "rhetor"—a discrete, rational being—to decode a "rhetorical situation" from outside (step one), and then consciously to select "appropriate" arguments (step two), *kairos* provides a point of departure from reasoned, linear steps—even from consciousness. As White observes, "The rhetorical practice of the sophist who allows *kairos* to figure in the invention of speech will issue, then, in an endlessly proliferating style deployed according to no overarching principle or rational design. The orator who invents on the basis of *kairos* must in fact always go beyond the bounds of the 'rational'" (21). Here White raises an important point about *kairos*'s relationship to subjective reason: the movements and betweenness of *kairos* necessitate a move away from a privileging of "design" or preformulated principles. At times, however, these so-called principles could be so habituated as to not require "thinking" *per se*. Janet Atwill presses on similar issues in her study of rhetorical *technê* (art), the embodiment of the art that constitutes a performance, practice-based "knowledge" of *kairos*. Atwill observes that "'knowing how' and 'knowing when' are at the heart of *kairos*, distinguishing *technê* from rule-governed activities that are less constrained by temporal conditions" (59). Kairotic impulses can therefore be habituated or intuitive—even bodily—and are not limited to a seat of reason or conscious adherence to a set of precepts. Rather, they depend largely on the rhetorical encounter itself and the forces pushing on the encounter. Such encounters mandate responses, and these responses can connect and hence lead to other emergent forces, while severing others.

Again, *kairos* serves as a useful conceptual tool with which to think about these cuttings. As White puts it, "*kairos* regards the present as unprecedented, as a moment of decision, a moment of crisis" (14). Here, the word "decision" demarcates the particular action of rhetoric. From the Latin verb *decidere*, which means "to cut off," a decision necessarily entails a cutting. The mythical figure of Kairos epitomizes decision- and incision-making in that he is usually depicted bearing scales and razor blades, tools for measuring and cutting as well as for being measured and cut. Figure 1, for example, shows Kairos perched on a narrow surface, balancing a pair of scales on the edge of a razor blade. His excellent balancing skills are key: Kairos must remain in the middle, ever ready for a moment of intervention. The god thus illustrates what I would call a "rhetorical stance"; he is on his toes, prepared for action, and attuned to the forces at work at a particular time. The god Kairos stands as a figure of in(ter)vention insofar as *kairos* mediates—or goes "between"—the outside of the self, i.e., the nodes where the "self" encounters the world, and the discourse or the "other" that the self encounters. In some depictions of Kairos, for example, the god is exerting pressure on the scale closest to him. This action might be read as a kind of material encounter with justice, where Kairos provides interventional tools with which to transform the outcome of a particular encounter. Because his razor blades help him intervene "in the nick of time," Kairos designates the moment of the encounter between self and other.

Again, the *Helen* provides a useful example of this betweenness, this interventional slicing/being sliced into a situation. Gorgias cuts into the discourses already circulating about Helen, deciding to use some existing discourse, while ignoring (i.e., selecting out) others. For example, Gorgias is compelled to take on "those who rebuke Helen, a woman about whom there is univocal and unanimous testimony among those who have believed the poets and whose ill-omened name has become a memorial of disasters" (2). As Kennedy points out in his discussion of the *Encomium of Helen,* however, Gorgias "ignores the more favorable treatments of Helen in Stesichorus' *Palinode,* Herdodotus' *Histories,* and Euripides' *Helen* (284, n. 1), seconding instead the general discourses that blame her for the Trojan War. Yet he contrasts this movement with another set of stories about Helen, the genealogical *mythoi,* which directly connect her to the immortals: "Now that by nature and birth the woman who is the subject of this speech was preeminent among pre-eminent men and women, this is not unclear, not even to a few; for it is clear that Leda was her mother, while as a father she had in fact a god, though allegedly a mortal, the latter Tyndareus, the former Zeus" (Gorgias 3). Gorgias goes on to connect her high birth with her "godlike beauty," an observation that echoes his enumeration of praiseworthy virtues in line 1: those attributes praiseworthy in a city, the soul, action, and speech, are a good army, wisdom, virtue, and truth, and a beautiful body deserve praise as well. He omits details about Paris, though he speaks of that omission: "for to tell the knowing what they know is believable but not enjoyable" (5).

Here, Gorgias cuts and is cut in between the divergent discourses—those proclaiming Helen's divine lineage and those associating her name with evil—in a manner similar to the *Dissoi Logoi*. This sophistic treatise establishes the in-betweenness of discourse by expounding on the various sides of several issues—justice, beauty, virtue, the good, and the seemly—and of course, their opposites: "take the case of various contests, athletic, musical, and military; in a race on the stadium, for instance, victory is good for the winner but bad for the losers. The same holds true for wrestlers and boxers, and for all those who take part in musical contests: for instance, victory in lyre-playing is good for the winner but bad for the losers" (DK 90.I.6–7; Sprague 280). Like the *Dissoi Logoi,* Gorgias' *Helen* moves from the already established views to different views, constructing a veritable chain of instances where the same act can be viewed as, for example, both good *and* bad. Gorgias offers several possible explanations for Helen's flight, each one adding doubt to her responsibility for her own actions. Such explanations include the forces of witchcraft, speech, and love. Gorgias does not settle on one definitive explanation, but enumerates several viable ones. In so doing, he resists pressing an ontological stamp on the situation: "Helen is not guilty because. . . ." Instead, he moves between the reasons she might not be to blame—thus creating a conjunction of forces ("and . . . and . . . and . . . and").[6]

The movement of Gorgias' speech, then, occurs in the middle, in the realm of the between. As Deleuze and Guattari put it, "*Between* things does not designate the localizable relation going from one thing to the other and back again, but a perpendicular direction, a transversal movement that sweeps one *and* the other away, a stream without beginning or end that undermines its banks and picks up speed in the middle" (25). In other words, Gorgias' betweenness, not necessarily Gorgias himself, seeks not to replace the previously accepted "truth" about Helen with another truth, but rather to undermine the very notion that one truth (or any truth for that matter) exists (much like the *Dissoi Logoi*). Sophistic truth resides in the logic of the "and"—in the connective force which constitutes the "turn" on which Gorgias impels his listeners to focus. Deleuze and Guattari explain the force of the "and":

> There has always been a struggle in language between the verb *etre* [*eimi*] (to be) and the conjunction *et* [*kai*] (and) between *est* and *et* (*is* and *and*). It is only in appearance that these two terms are in accord and combine, for the first acts in language as a constant and forms the diatonic scale of language, while the second places everything in variation, constituting the lines of a generalized chromaticism. From one to another, everything shifts. . . . That is when style becomes a language. That is when language becomes intensive, a pure continuum of values and intensities. (98)

In other words, the "and" interrogates sweeping ontological claims by pointing to the multiplicity that emerges on the other end of the "is": "x is y *and* z," etc. The "and," a copulative word, thus exposes the limit of the "is," which Deleuze and Guattari compare to the standard diatonic, eight-tone musical scale, and thus assures variations or turns, enabling thought to depart and go somewhere else.

In a key moment at the end of the *Helen*, where Gorgias proclaims he has effectively "removed disgrace from a woman," he also adds: "I wished to write a speech which would be a praise of Helen *and* a diversion (*paignion*) to myself." This line, one of the most intensely studied lines of Gorgias', produces significant effects. On one hand, the two-pronged claim troubles generic constraints. Insofar as the *Helen* presses on what counts as "truth," this final statement functions to upend the traditional genre of epideictic, going beyond the generic constraints of praise and blame to suggest that the speech has also served a recreational function.[7] By extension, the final comment renders the entire speech a *paignion*, a game, and therefore undermines notions of "truth" in general; in short, the gaming aspect trumps the persuasive function of the *Helen,* i.e., the actual removal of blame. Poulakos offers this reading, writing that Gorgias "is content to have participated in the game of words, to have demonstrated to his audience that he is a splendid player, and to have tried to bring them into the game" (67). This final phrase (*emon de paignion*) folds truth back into the notion of a game, a *paignion,* hence subordinating a truth discourse to a playful jest.

On the other hand, Gorgias' last sentence sports the logic of the "and," wherein the two functions of the speech are not hierarchized but maintained as dual directions. The *Helen* is thus not one singular thing, but rather runs between—at once a discourse which challenges established truths about Helen and also a recreational discursive movement. The "and" thus effectively produces tension with the singular "would be" that precedes it ("I wished to write a speech that *would be* . . ."), splitting the notion of being and hooking it into other objects: *both* Helen *and* Gorgias himself. The "and" therefore serves the transformative function of "invention-in-the-middle," dispersing force between and among Helen, Gorgias, and—lest we forget—those who have been bearing the "turns" along the way.

The Gorgianic *Logos-Dunamis* Complex

Perhaps this Gorgianic emphasis on "turning" and its relationship to subjectivity can be illuminated by linking them to Gorgias' thought on the movement of fluids through the body. In Plato's *Meno,* Gorgias is said to have followed Empedocles in his belief that "existing things have some effluences . . . and pores into which and through which the effluences are carried" (76C). The effluences, fluids, films, or smells emitted or transmitted through the body were thought to be suited to the sensory perceptions. Further, Gorgias was said to have thought that fire moved through pores of materials in a similar manner, as evidenced in Theophrastus' refutation of Gorgias' theory that combustion from mirrors and other shiny surfaces takes place "'by means of the fire passing away through the pores'" (DK 82B5; Sprague 47). For Gorgias, bodies and souls, like bronze and silver, were porous entities which allowed effluents and other substances (words, fire) to pass through. No wonder his speeches were referred to as "torches" (Smith 359).This somatic relation also emerges in the

Helen when Gorgias suggests an analogy between speech and drugs: "The power of speech has the same relation to the disposition of the soul as the application of drugs on the disposition of the body. For just as different drugs draw different juices out of the body, and some end disease but others end life, so also some speeches produce pain, some enjoyment, sòme fear; some instill courage in hearers; some drug and beguile the soul with a kind of evil persuasion" (14).[8] So for Gorgias, speech (*logos*) can move through the soul (*psyche*) like drugs through the bloodstream, and the effects can be as potent as hemlock or as soothing as rubbing oil. Just as Gorgias focused on the turn in the passage discussed earlier, here he focuses on the changes produced on the disposition (*taxis*) of the body or soul when drugs and speech pass through, thus emphasizing the *encounter* between body and drug, speech and soul. Toward the end of this passage, Gorgias uses the same root term that he uses for drugs—*pharmakôn*— in verb form (*pharmakeusan*) to suggest what speech does: some *logoi* "drug and beguile the soul." At this point it is possible that drugs are no longer metaphorical for *logoi,* but that *logos* becomes a type of *pharmakon.* Gorgias' *pharmakon* resembles the *pharmakon* in Jacques Derrida's reading of the concept in Plato. Derrida points out that "The *pharmakon* is that which, always springing up from without, acting like the outside itself, will never have any definable virtue of its own" (102). That is, a *pharmakon* (translated variously, as Derrida points out, as "drug," "remedy," "poison"), can only be considered in relation to something else, some other body, and its effects on a particular body cannot be known in advance. As Derrida puts it, "in order for this *pharmakon* to show itself, with use, to be injurious, its effectiveness, its power, its *dunamis* must, of course, be ambiguous" (103).

Not unlike Derrida, Gorgias acknowledges the most prominent characteristic of *logos:* its *dunamis,* its capacity to effect change. The *dunamis* of *logos* therefore cannot be known in and of itself—i.e., *dunamis* does not take a "form," but emerges only relationally—between drugs and blood, even between wrestlers. The French word for power, *pouvoir,* aligns with *dunamis* in interesting ways. As Gayatri Spivak points out, "*Pouvoir* is of course 'power.' But there is also a sense of 'can-do'-ness in *pouvoir,* if only because, in its various declinations it is the commonest way of saying 'can' in the French language" (qtd. in Biesecker 355). Bearing in mind that *pouvoir* indicates "capacity" and "potential" helps to elucidate Foucault's suggestion that "power produces; it produces reality; it produces domains of objects and rituals of truth" (*Discipline and Punish*

194). *Dunamis* also captures the various forces of *pouvoir,* for it can be translated as "power," "influence," or "forces," "means," and "function," "faculty," "capability." *Dunamis,* then, can also be said to produce reality. Linking *dunamis* to a productive notion of power yields a different reading of Gorgias' notion of discourse. Gorgias' explication of the *logos-dunamis* complex might suggest something other than speech as a means of control or domination (persuasion). Instead, Gorgias' comparison of the power of speech (*ê te tou logou dunamis*) to drugs (14) might be read as a suggestion of the "'can-do'-ness" of speech. The function of *logos,* then, resides in its relations, at the specific junctures where it encounters and is encountered by other forces.

A consideration of language or rhetoric as an encounter of forces taps into a set of philosophical insights about the performativity of language as outlined by J. L. Austin and taken up by Derrida, Judith Butler, and others.[9] Drawing on an Austinean-Derridean account of performativity, Richard Doyle offers an important insight about the ways in which an account of language-as-force troubles notions of hermeneutical or representational accounts of communication. For Doyle, "the transmission, passage, and communicability of language [. . .] become something other than an affair of meaning or information; they become something more like ballistics or contagion, the transmission and repetition of an effect across bodies of discourse and across bodies" (5). Such performative effects of language align with Gorgias' notion of language's drug-like effects, and its capacity or *dunamis* to incite movement.

The issue of a broad *dunamis* (i.e., the possibility of producing multiple effects) is never far from the ethical issue of use. The general point in section 14 of the *Helen* is similar to that made in Plato's *Gorgias* (456a–457) when Socrates, marveling about the power (*dunamis*) of rhetoric, worries about the ethical ramifications of teaching such a forcible art. In response to Socrates, Gorgias argues that the teachers should not be blamed for the misuse of rhetoric, this time making a comparison to athletic training: "just because one has learned boxing or wrestling or fighting in armour so well as to vanquish friend and foe alike: this gives one no right to strike one's friends, or stab them to death" (456d). Here, as in the *Helen,* Gorgias acknowledges the most prominent characteristic of *logos:* its *dunamis,* its capacity to (effect) change, and concomitantly, its potential (use for) destructive ends. By establishing that this broad *dunamis* is a characteristic of *logos* in general, Gorgias' discourse attempts to free from future blame the teachers of rhetoric, just as the speech frees Helen from blame. But both discourses maintain a key point for sophistic movement: the *dunamis* of *logos,* like the bodily arts of pharmacology and athletic training, emerges in the encounter itself.

Gorgias' arguments about the power of *logos* underscore the kairotic tenor of his rhetoric. Specifically, the art of pharmacology creates another link to *kairos.* In antiquity, the practice of medicine was guided by a logic of *kairos.* Aristotle regarded medicine as the most practical and common sphere for the use of *kairos,* as he writes in the *Ethics* of the importance of circumstance: "And if this is true of the general theory of ethics, still less is exact precision possible in dealing with particular cases of conduct; for these come under no science or professional tradition, but the agents themselves have to consider what is suited to the circumstances on each occasion *(kairon),* just as is the case with the art of medicine" (2.2.4). The production of a bodily state—be it the emission of particular fluids, the relief of pain, or the production of pleasure—depends on the singular encounter between the drug and the body in a particular condition at a particular moment. Aristotle wasn't the only one who viewed the practices of medicine as a kairotic encounter. As G. E. R. Lloyd points out, the Hippocratic tradition held that diseases can be successfully treated "if you hit upon the right moment *(kairos)* to apply your remedies" (362; also qtd. in Atwill 57). Similarly, the effects of athletic practices of boxing and wrestling cannot be known in advance, but rather depend on a particular encounter —the *agôn*—which demands a deployment of skills on the spot, in the heat of the moment, in the blink of an eye. For Gorgias, both athletics and pharmacology capture the kairotic capacity of discursive practices.

This relational specificity—the emergence of effects from specific encounters—helps account for both the importance of *kairos* in sophistical rhetoric and for the general dissatisfaction with Gorgias' attempts to explain *kairos.* Dionysus of Halicarnassus, for example, complains: "No orator or philosopher has up to his time denied the art of the 'timely,' not even Gorgias of Leontini, who first tried to write about it, nor did he write anything worth mentioning" (Sprague 82B13). Untersteiner and Rostagni both contend that Dionysius only considered "pedantic formal classifications" to be worth mentioning (Untersteiner 203, n. 11). However, enumerating the precepts of *kairos* would prove counter to Gorgias' rhetoric —and to *kairos* itself. It is precisely because of this relational specificity that Gorgias "seems" to have no guiding "theory" of rhetoric or its uses. In many ways, Plato's Socrates poses the impossible ethical question to Gorgias, asking him to account for rhetoric's capacity for "misuse."

If sophistic rhetoric depends on the specific encounter, then *kairos* is Gorgias' ethics.

So far, I have tried to address problems with the "subject" of invention and to rearticulate the "discovery" and "creation" models of rhetorical invention to allow for postmodern critiques of the singular sovereign subject. I have suggested "invention-in-the-middle" as an alternative mode of rhetorical invention, one that depends on a reshaping of rhetoric itself. This particular reshaping invokes the movement of discourse, rhetoric's betweenness, and the productive dimension *logos*'s power (*dunamis*). While this view of rhetoric certainly troubles programmatic approaches to discourse production and, consequently, to rhetorical invention, I would argue that this reconceptualization of invention is tenable. When rhetoric emerges from encounters, invention is practiced on many levels: in the unexpected syncopations that occur under the traditional rubric of "style"; in the strategic piecing together of discourse (also called "arrangement"); in the bodily and "surface" movements sometimes called "delivery"; and in the configurations of experiential "memory," which necessarily entails the act of forgetting.

In other words, the traditional canons of rhetoric become less discrete, useful only for naming actions—for distinguishing the cuttings and connections, the "contagion"-like "transmissions" effected by discourse's *dunamis*. To be sure, the logic of the "and" replaces the linear "then." Rather than the five-step program ("invention, then style, then arrangement . . ."), the canons would cluster around "ands," held in tension, and enacted only through the movements—or turns—of discourse. As I have tried to suggest, these canonical categories have little meaning without the relational specificity of particular encounters. It is only through the timely, *kairotic* encounter that "turns" happen, different *ethoi* emerge, and *logos* becomes action—or, in a reversal of Electra's middle statement—words make themselves deeds.

Notes

1. See White's chapter on middle voice, which wonderfully situates the middle voice in Freudian work on sadomasochism. White's elaboration of and departures from the psychoanalytic tradition performatively produce the mobile fluctuating discourse which his book attempts to elaborate.

2. The translations of fragments about Gorgias' life are taken from George Kennedy's contribution to Rosamand Kent Sprague's edition of *The Older Sophists*.

3. For the usefulness of *kairos* in contemporary pedagogy, see James Kinneavy and Michael Carter. James S. Baumlin offers an in-depth discussion of the concept's relationship to *prepon* or decorum. More philological treatments of *kairos* include those of William H. Race, Richard Broxton Onians (esp. pages 343–48), and J. R. Wilson (1980 and 1981). For important and rarely cited (at least in rhetorical scholarship) philosophical work on *kairos*'s relationship to *chronos*, see John E. Smith (1969 and 1986). Also, Carolyn Miller (1992 and 1994) has made good use of the concept in relation to technology and the rhetoric of science.

4. For an excellent discussion about the bodies of mythic figures, see Jean-Pierre Vernant's "Dim Body, Dazzling Body."

5. Unless otherwise noted, all translations of Gorgias' *Helen* are taken from George Kennedy's version in his edition of Aristotle's *On Rhetoric* (284–88).

6. For the *and*'s relationship to the conjunctive fabric of the rhizome, Deleuze and Guattari's theoretical tool for opposing the ontology of the tree, see *A Thousand Plateaus* (25).

7. Isocrates, in his *Encomium on Helen*, chides Gorgias for violating the conventions of the encomiastic genre: "Nevertheless, even he committed a slight inadvertence—for although he asserts that he has written an encomium of Helen, it turns out that he has actually spoken a defense of her conduct!" (14–15).

8. Translation mine.

9. Austin's theory of performativity explores the notion that language performs actions on subjects, as in the marriage ceremony's "I now pronounce you husband and wife."

Works Cited

Aristotle. *Nicomachean Ethics*. Trans. H. Rackham. Cambridge: Harvard UP, 1934.

Atwill, Janet. *Rhetoric Reclaimed: Aristotle and the Liberal Arts Tradition*. Ithaca: Cornell UP, 1998.

Baumlin, James S. "Decorum, *Kairos*, and the 'New' Rhetoric." *Pre/Text* 5 (1984): 171–83.

Biesecker, Barbara. "Michel Foucault and the Question of Rhetoric." *Philosophy and Rhetoric* 25 (1992): 351–64.

Butler, Judith. *Excitable Speech: A Politics of the Performative*. New York: Routledge, 1997.

Carter, Michael. "*Stasis* and *Kairos:* Principles of Social Construction in Classical Rhetoric." *Rhetoric Review* 7 (1988): 97–112.

Cook, Arthur Bernard. *Zeus: A Study in Ancient Religion.* New York: Biblo and Tannen, 1965.

Deleuze, Gilles. *Foucault.* Minneapolis: U of Minnesota P, 1988.

Deleuze, Gilles, and Félix Guattari. *A Thousand Plateaus: Capitalism and Schizophrenia.* Minneapolis: U of Minnesota P, 1987.

Derrida, Jacques. *Dissemination.* Chicago: U of Chicago P, 1981.

Diels, Walther, and Hermann Kranz. *Die Fragmente der Vorsokratiker.* Berlin: Weidmann, 1951–52.

Doyle, Richard. *On Beyond Living: Rhetorical Transformations of the Life Sciences.* Stanford: Stanford UP, 1997.

Faigley, Lester. *Fragments of Rationality: Postmodernity and the Subject of Composition.* Pittsburgh: U of Pittsburgh P, 1992.

Foucault, Michel. *The History of Sexuality.* New York: Random, 1978.

———. *Discipline and Punish: The Birth of the Prison.* New York: Vintage, 1979.

———. *The Use of Pleasure: The History of Sexuality,* Vol. 2. New York: Vintage, 1990.

Gorgias. *Helen.* Trans. George Kennedy. Aristotle. *On Rhetoric: A Theory of Civic Discourse.* Trans. George Kennedy. New York: Oxford UP, 1991. 284–88.

Isocrates. *Helen. Isocrates.* Vol. 3. Trans. Larue Van Hook. Loeb Classical Library. Cambridge: Harvard UP, 1961. 54–97.

Kinneavy, James L. "*Kairos:* A Neglected Concept in Classical Rhetoric." *Rhetoric and Praxis.* Ed. Jean Dietz Moss. Washington, D.C.: Catholic U of America P, 1986.

Liddell, Henry George, and Robert Scott. *A Greek-English Lexicon.* Oxford: Clarendon, 1996.

Lloyd, G. E. R. *Methods and Problems in Greek Science.* Cambridge: Cambridge UP, 1991.

Miller, Carolyn. "Opportunity, Opportunism, and Progress: *Kairos* in the Rhetoric of Technology." *Argumentation* 8 (1994): 81–96.

———. "*Kairos* in the Rhetoric of Science." *A Rhetoric of Doing: Essays Honoring James L. Kinneavy.* Ed. Steven P. Witte, Neil Nakadate, and Roger Cherry. Carbondale: Southern Illinois UP, 310–27.

Muckelbauer, John. "Re: Producing Resistance: In Search of M. Foucault." Penn State Conference on Rhetoric and Composition. State College Pennsylvania, 7 July 1997.

Onians, Richard Broxton. *The Origins of European Thought: About the Body, the Mind, the Soul, the World, Time, and Fate.* Cambridge: Cambridge UP, 1951.

Plato. *Meno. Plato II: Laches, Protagoras, Meno, Euthydemus.* Cambridge, Harvard UP, 1990.

Plato. *Gorgias.* Cambridge: Harvard UP, 1991.

Poulakos, John. *Sophistical Rhetoric in Classical Greece.* Columbia: U of South Carolina P, 1995.

Race, William H. "*Kairos* in Greek Drama." *Transactions of the American Philological Association* 111 (1981): 197–213.

Schiappa, Edward. "An Examination and Exculpation of the Composition Style of Gorgias of Leontini." *Pre/Text* 12 (1991).

Segal, Charles P. "Gorgias and the Psychology of the Logos." *Harvard Studies in Classical Philology* 66 (1972): 99–155.

Smith, Bromley. "Gorgias: A Study of Oratorical Style." *Quarterly Journal of Speech Education* 7 (1921): 335–39.

Smith, John E. "Time, Times, and the 'Right Time': *Chronos* and *Kairos.*" *The Monist* 53 (1969): 1–13.

———. "Time and Qualitative Time." *Review of Metaphysics* 40 (1986): 3–16.

Sophocles. *Electra.* Teubner: Stutgardiae, B.G., 1996.

Sprague, Rosamond Kent, ed. *The Older Sophists: A Complete Translation by Several Hands of the Fragments in Die Fragmente Der Vorsokratiker.* Columbia: U of South Carolina P, 1972

Untersteiner, Mario. *The Sophists.* Trans. Kathleen Freeman. New York: Oxford UP, 1954.

Vernant, Jean-Pierre. "Dim Body, Dazzling Body." *Fragments for a History of the Human Body.* Ed. M. Feher. New York: Urzone, 1989.

White, Eric Charles. *Kaironomia: On the Will-to-Invent.* Ithaca: Cornell UP, 1987.

Whitney, Geoffrey. *A Choice of Emblems 1586.* Ed. John Horden. Yorkshire: Scholar, 1973.

Wilson, J. R. "KAIROS as 'Due Measure.'" *Glotta* 58 (1980): 177–204.

———. "KAIROS as 'Profit.'" *Classical Quarterly* 31 (1981): 418–20.

Young, Richard, and Yameng Lieu. Introduction. *Landmark Essays on Rhetorical Invention.* Ed. Richard E. Young and Yameng Liu. Mahwah: Erlbaum, 1994.

Rhetoric and Hermeneutics

Division through the Concept of Invention

ARABELLA LYON

> Rhetoric, that practically conscientious discourse of struggle and conflict, has been aestheticized. . . . The cult of textuality has had the effect of blinding many of us to and also insulating many of us from the places where real material grievances are stored and sometimes lost.
>
> *Thomas B. Farrell, "Disappearance" (1993)*

> I did not translate them as an interpreter but as an orator.
>
> *Cicero,* De optimo genere oratorum *(46 B.C.E.)*

Many rhetoricians, especially those concerned with postmodern theory, neglect the complexities of rhetorical production, replacing rhetoric—historically concerned with intersubjective issues of power, manipulation, persuasion—with textuality.[1] With this maneuver, negotiation and deliberation between a rhetor and an audience become only a slim piece of the discussion, and the canons of rhetoric (invention, arrangement, style, memory, and delivery) all but disappear. Rhetoric becomes a strategic approach to reading texts, not a *praxis* or *poesis*.

By turning toward interpretation and away from production and making, rhetoricians have diminished the place of rhetoric as an action in the world. Rhetorical action, while dependent on the entire canon of rhetoric, begins with invention; the possibility of action, rather than simple motion, is dependent on the invention of an act, whether through topics, chance, achieving identity, or assessing landscapes. Rhetorical invention initiates and constantly intervenes in rhetorical production. As I will demonstrate, at the other extreme, interpretation, and so interpretive invention, consists of mediating between a reader (whether an interpreter or an audience) and an extant text. Rhetoric and interpretation involve different inventional strategies.

S. Michael Halloran warned that the differences between the classical and modern periods would affect our understanding of theories of invention. As we begin the twenty-first century, the ancient concept has only become more problematic, especially postmodern theory's suspicion of agency, origins, form, and progress, and its emphasis on multiple discourses, deconstruction, irony, fragmentation. In this frame, invention becomes a term in need of reclamation and redefinition.

For purposes of understanding rhetorical actions and their necessary relationship to invention, this essay examines the relationship between rhetoric and hermeneutics, a particular kind of interpretation, for three reasons. I focus here, in part, because rhetorical reading is the unstable bridge between rhetoric and hermeneutics which helps to define invention: the nature of invention becomes visible as far as we are able to divide hermeneutics from rhetoric. I also look at hermeneutics and rhetoric because through this juncture I am able to see the inventional strategies of a rhetor without losing awareness of an audience; that is, the role of identification and the resisting reader, especially important in a postmodern rhetoric, becomes visible. Finally, a discussion of invention in the context of the relationship between hermeneutics and rhetoric demonstrates what is lost if rhetoric, and so rhetorical invention, are not beheld as "making" the discursive world.

Placing Debates about Rhetoric and Interpretation

Ultimately, hermeneutics is useful to rhetoric, but it is not the same project. Concerned with texts and their relationship to historical consciousness, hermeneutics is not alien to rhetoric, nor is either activity autonomous. However, a too ready collapse of one into the other destroys their differences, differences necessary to understanding rhetoric as deliberative and inventive. It is commonly acknowledged that rhetoric is easily transfigured into other activities,[2] but as Thomas B. Farrell writes (of epistemic rhetoric in particular), "[a]ll knowledge is rhetorical in direct proportions to how trivial is one's initial conception of rhetoric" ("Parthenon" 82).[3] True, this is a rather snide comment, but it identifies the core problem of institutionalized rhetoric at the end of the twentieth century. That is, rhetoric represents too many things to too many scholars and winds up wearing too many caps. The winged cap of Hermes is one of them.

While rhetoric and hermeneutics have overlapped from the classical period onward, they have become increasingly allied, with neither the purpose nor effect of the alliance adequately analyzed.[4] Fueled by increased studies in rhetorical criticism and rhetorical hermeneutics and by the disciplinary move of rhetoric from departments of communication to departments of literatures, this alliance suggests the evolution of a new rhetoric removed from the practice of textual production within public spheres. There are methodological and institutional benefits to this reconfiguration and relocation of rhetoric. Hence, some even welcome this change to an interpretive rhetoric, but others from humanist, feminist, compositionist perspectives try to modify the discipline's turn. Dilip Gaonkar and Michael Leff respectively represent two positions on the new, new rhetoric.

In "The Idea of Rhetoric in Rhetoric of Science," Gaonkar analyzes the causes and effects of the interpretive turn, arguing that the universalization of the concept of rhetoric, its transformation to "a hermeneutic metadiscourse" from a once "substantive discourse practice" (258), is a response to the epistemic crisis caused by the demise of foundationalism. He asserts that this transformation of rhetoric serves to deflect our attention from "finding what motivates and steers rhetoric" (267). Given our distance from traditional rhetoric and our fatigue (I would add confusion) over the promiscuous use of the term rhetoric, Gaonkar would have rhetoric function globally as a supplement added to other discourses and interpretive instruments. However, in what is a complex procedure, he does not attend to "what motivates and steers rhetoric," at least historically. Instead he promotes the hermeneutical turn, resisting a return to the rhetorical tradition because of its emphasis on an outdated "ideology of human agency" (275).

Leff's humanist response critiques Gaonkar's approach as transferring agency from the rhetor to the audience and "from the forum, the lawcourt, and the pulpit" to "the study, the lecture hall, and the library" ("Idea" 298). That is, Leff sees Gaonkar's stance on agency, the basis of his claim of distance from the tradition, as shuffling agency from one human to another, from author to audience, with no real evolution or critique of our understanding of agency. Leff defines the interpretive turn in rhetoric as a turn from the public sphere to an aestheticized, academic realm and as a turn from the agency of the rhetor to the agency of the audience. I will later argue that this move from the agency of the rhetor to the agency of the audience displaces invention; in fact, it may even destroy any place for theorizing rhetorical invention.

While Gaonkar and Leff draw attention to the relationship between interpretation and agency in rhetoric, neither significantly advances present debates. They demarcate the debate's territorial boundaries, but they do not suggest a new practice of—or a route for—either returning to an older, political rhetoric or adapting hermeneutics as an instrument for understanding the productivity of rhetoric.

Are Rhetoric and Hermeneutics Synonymous?

Gaonkar and Leff leave us with a dichotomy between interpretation and traditional rhetoric and little sense of how a theory of rhetorical invention, one that addresses concerns of postmodernity, might function. That is, they problematize rhetorical production, but give us no solutions to the dearth of theory on rhetoric as a public practice. I believe the concept of invention allows us to begin to separate hermeneutics from rhetoric; moreover, this process of differentiation shows both where rhetorical invention lies and how inventive rhetorical invention can be. Before demonstrating the different functions of invention in rhetoric as the drive to action and hermeneutics as the contemplation of meaning, I would like to review two major discussions of the rhetoric/hermeneutic relationship, Hans-Georg Gadamer's and Stephen Mailloux's. Gadamer sees rhetoric and hermeneutics as synergistic, and Mailloux sees them as synonymous. Many have premised the connection, but I will argue that the tension posited between rhetorical hermeneutics and hermeneutical rhetoric exists within the *interpretive frame* of hermeneutics, not within the *productive frame* of rhetoric. The relationship is only analyzed from a hermeneutical perspective, and this makes the relationship an unhealthy one for rhetoric.

Because Leff, Gaonkar, and so many others acknowledge Gadamer's hermeneutics as a model with positive implications for interpretive rhetorics, it would seem that Gadamer's hermeneutics, his ontological focus, and his definition of its relationship to rhetoric is the place to start analyzing interpretation, agency, and invention. For Gadamer, hermeneutics is never equivalent to rhetoric or to persuasion, as the negotiation of different discourses, nor is pedagogy. Hermeneutics as an art and rhetoric as a faculty are significantly different. He defines hermeneutics as "the art of clarifying and mediating by our own effort of interpretation what is said by persons we encounter in tradition" (*Philosophical* 98), and he defines rhetoric a "task" for philosophers: "to master the faculty of speaking in such an effectively persuasive way that the arguments brought forward are always appropriate to the specific receptivity of the souls to which they are directed" (21). According to Gadamer, these *two separate studies,* as well as sociology, are interdependent because of the universality of linguisticality, but they work synergistically.

Gadamer's hermeneutical orientation and concern with philosophy rather than political action leave him with a limited sense of rhetoric.[5] Because his theory of rhetoric is tied too closely to his Platonic concerns with interpretive truth and souls, it is not a great aid to understanding invention, deliberation, instrumental uses of discourse, and social and discursive power. These are not his projects; he is, by his own acknowledgment, promoting a model of interpretive truth with minimal attention to many familiar rhetorical concerns, dismissing both "all the shallow claims put forward by the contemporary teachers of rhetoric" and modern rhetoric's concern with "organizing a perfect and perfectly manipulated information" (21, 25). He fears persuasion.

Still, Gadamer sees rhetoric as helpful to hermeneutics. In addition to their interdependence through language, Gadamer seeks to ally hermeneutics with rhetoric because of its priority, vigor, and practical nature. He writes:

> Thus the rhetorical and hermeneutical aspects of human linguisticality interpenetrate each other at every point. There would be no speaker and no such thing as rhetoric if understanding were not the lifeblood of human relationships. There would be no hermeneutical task if there were no loss of agreement between the parties of a "conversation" and no need to seek understanding. The connection between hermeneutics and rhetoric ought to serve, then, to dispel the notion that hermeneutics is somehow restricted to the aesthetic-humanistic tradition alone and that hermeneutical philosophy has to do with a "life of the mind" which is somehow opposed to the world of "real" life and propagates itself only in and through the "cultural tradition." (*Reader* 280; see also *Philosophical Hermeneutics* 25)

His quotations around "conversation," together with Gadamer's Platonic sense of a philosophical, truth-serving rhetoric, are telling. As suggested earlier, this model does not suit rhetoric in its forensic or deliberative manifestations. In forensic and deliberative rhetorics, the "loss of agreement between the parties of a 'conversation'" may have little to do with a loss of meaning and understanding and far more to do with diverging political and social purposes. But in *contradiction,* what Gadamer wants from rhetoric is "'real' life" to allow hermeneutical philosophy to escape the library and the parlor. He wants the conversation and understanding together with "'real' life" without a wrestle in the scramble, lie, and injury. To even suggest that hermeneutics has "'real' life" makes hermeneutics dependent on rhetoric which must then precede and teach hermeneutics. He writes, "hermeneutics may be precisely defined as the art of bringing what is said or written to speech again. What kind of an art this is, then, we can learn from rhetoric" (*Reason in the Age of Science* 119; see also *Philosophical Hermeneutics* 21–26). The "again" is key here. Hermeneutics is a re-vision of an earlier production, an earlier invention. Hence, hermeneutics is dependent on what is said or written. There is a crucial rhetorical event (invention) prior to interpretation.

On the one hand, Gadamer's differentiation of rhetoric and hermeneutics proves useful both in suggesting the limited and ineffectual nature of the institutionalized version of each and in showing how a definition of one limits the terms of the other. On the other hand, Mailloux's rhetorical hermeneutics, so bare of differentiation as to make it simply a repetition of synonyms, emphasizes the difficulties of understanding invention within the collapse, and it serves to show how the work of rhetorical invention is different than that of hermeneutics.

Despite his rich, careful readings of nineteenth-century debates about *Huckleberry Finn* and "Star Wars" politics in *Rhetorical Power,* Mailloux's theory is not particularly helpful in regaining a deliberative sense of rhetoric. His is a case where a rich practice based in political debates and startlingly sharp reads of controversial texts remains tied to the theories of literary criticism—despite his affiliations with rhetorical theory. In delineating and accentuating interpretive aspects of rhetoric, he deletes the rhetorical tradition of textual production, a move that is surprising in light of his definition of rhetoric as "the political effectivity of trope and argument in culture" ("Articulation and Understanding" 379). Despite its seeming concern with rhetorical effect in politics and argumentation, this definition emphasizes the cultural effect of and audience response to a text and ignores the rhetor's activity of purposeful production. Mailloux's interest in studying historically based arguments ultimately evades the rhetorical tradition of teaching individuals "the available means of persuasion." If the rhetorical theorist is always the observer or interpreter of rhetorical exchanges, in this model the observations are of text and cultural effect, how a text is interpreted, and not the traditional rhetorical observations of what the rhetor can do to *produce* a desired effect, a specific action. This small shift has significant implications.[6]

Let me demonstrate the difficulty in his linking rhetoric and hermeneutics. In "Articulation and Understanding," Mailloux outlines a collapsed relationship between the two by defining them as synonyms. I would suggest that at points he is too facile in allying rhetoric and hermeneutics and, correspondingly, too concerned with interpretation and insufficiently concerned with production to define a rhetoric that allows for deliberation and action. He begins his definitions by writing that the "traditions of rhetoric and hermeneutics address very practical tasks. Hermeneutics deals with interpretation focused on texts, and rhetoric deals with figuration and persuasion directed at audiences. *Interpretation* can be defined as the establishment of textual meaning, while *rhetoric* (as figuration and persuasion) might be characterized more pointedly as the political effectivity of trope and argument in culture" (379, emphasis added). So far so good, his emphasis on trope as particularly significant to rhetoric is odd, but not outside of Ramistic, belletristic, or French traditions.

Mailloux continues: "Interpretation involves the *translation* of one text into another, a Hermes-like mediation that is also a *transformation* of one linguistic event into *another* later one. Rhetoric involves the *transformation* of one audience into *another*, which is also a psychogogic *translation* from one position to a different one" (379, emphasis added). Here we have more trouble. I have underlined "transformation," "translation," and "another" because the words are clearly shifting dramatically in meaning as their context shifts. As a matter of textual play, this is pretty, but as a logical system, it falters.

The first "translation" is the dictionary definition "interpret, to put into different words or different language" (*Webster's Unabridged Dictionary* 1972). This definition is supported by Mailloux's appeals to "Hermes" and "mediation." That is, if translation is not a simple worldly process, it is about messages and bridging or bringing meanings. The second "translation" corresponds not so much to words as to positionality; it corresponds to a dictionary definition coupled with "transfer" (or "bear across"), such as "to convey to heaven," "to transfer (a bishop) from one see to another," "to move a saint's body or relic." By his own word choice, the translation of the audience is physical and spiritual, and so, involves a different realm of action than the discursive moves of interpretative translation.

The meanings of "transformation" into "another" do not so dramatically change their definitions by the sentence context as does "translation," but the issue is what is transformed and by how much. The "discursive event" is transformed into "another later one," and probably a different one, but that is not a requirement of his statement. Given the pun (purposeful?) on transformative grammar and the strong sense of "tradition" from which Mailloux is working, it is unclear as to how much the discourse can be transformed (and by whom); it is clear, however, that the text is passive. As required by the parallel structure, the audience transformation is also passive, according to Mailloux. That is, the audience, potentially many people, can be made into another group of people. This equation of persuasion with immediate psychological change is appealing (as are its implications for work with deans, legislators, and tenure committees), but any real discussion of transformations of audiences needs more nuance. Mailloux's premise of the passive audience reveals the degree of his commitment to avoiding issues of agency, deliberation, and the public sphere. Theories of passive audiences reveal little about deliberative processes and have been problematized by rhetoricians since Gorgias played at defending the passive Helen.

The definition intermingling rhetoric and hermeneutics continues with fewer difficulties, but it still continues to collapse the two activities and to erase key differences: "As practices, rhetoric and interpretation denote both productive and receptive activities. That is, interpretation refers to the presentations of a text in speech—as in oral performance—and the understanding or exegesis of a written text; similarly, rhetoric refers to the production of persuasive discourse and the analysis of a text's effects on an audience" (379). My point here, that rhetoric and hermeneutics both engage processes of production and reception, is not controversial. The controversies turn on the extent to which each is characterized by production and reception and the degree to which any type of production or reception is similar in the context of rhetor and audience purposes. So, is "oral performance" (a recitation or production of *Hamlet*) really the same "interpretation" that he means by "hermeneutics," or is "oral performance" only a somewhat related concept that is particular to dramatic, "presentational" aspects of meaning-making? Even if we relate presentation (memory and delivery) to rhetoric, the tradition of rhetoric describes five parts (invention, arrangement, style, memory, delivery), and so the production that he associates with interpretation is a thin one, scant of rhetorical invention, arrangement, and style, though full of memory and delivery. Mailloux is accurate in describing rhetoricians as analyzing the effectiveness of their discourse, but is this the same reception that hermeneutics describes? Hermeneutics is a theory not about the effect on an audience, but about the truth-seeking approach of an educated interpreter. Furthermore, rhetoricians analyze and understand effects on audiences and audiences themselves with their specific intentions and purposes. Among those purposes there is always an instrumental one, a pointing to further persuasions, conceivably mutual, and to further productions. Still, I can understand a text, especially a literary text, and have no purpose beyond pleasure or dalliance, no public task, at least.

Mailloux continues: "*In some ways* rhetoric and interpretation are practical forms of the same extended human activity: Rhetoric is based on interpretation; interpretation is communicated through rhetoric" (379, emphasis added). A true, but trivial connection; linguistic behaviors are a large, but finite, related set of actions. I argue that the ways in which rhetoric and hermeneutics differ as human activities are more revealing of language use and abuse.

Mailloux then concludes this section of the essay, writing:

> Furthermore, as reflections on practice, hermeneutics and rhetorical theory are mutually-defining fields: hermeneutics is the rhetoric of establishing meaning, and rhetoric is the hermeneutics of problematic linguistic situations. When we ask about the meaning of a text, we receive an interpretive argument; when we seek the means of persuasion, we interpret the situation. As theoretical practices, hermeneutics involves placing a text in a meaningful context, and rhetoric requires the contextualization of a text's effects. (379)

He has shifted here from the word "interpretation" to "hermeneutics," from "rhetoric" to "rhetorical theory." The earlier connections of practice, which I maintain are problematic, are now extended into theory and the fields' disciplinary manifestations.

There is a large literature, dating back to Schleiermacher, which stakes a claim that all interpretation is the province of the theoretical field of hermeneutics. There is also a large literature which stakes a claim that all texts, even all symbol systems, are rhetorical. Yet in moving to theory, Mailloux has come to question the relationship between rhetoric and hermeneutics: lots of interpretation happens without the "we" requesting an interpretive argument; lots of symbolizing gets done without the "we" interpreting the situation with any consciousness or efficacy. Hermeneutics may require argument; interpretation does not. Furthermore, saying that any use of argument is rhetorical is not the same as saying hermeneutics is uniquely or particularly related to rhetoric. With that line of thinking, any argument on any topic makes the topic itself into rhetoric. An argument about dinner does not make dinner into rhetoric; an argument about meaning does not make meaning into rhetoric.

One may concede that hermeneutics is argument is rhetoric; still, rhetoric, not necessarily rhetorical theory, is always more than "the hermeneutics of problematic linguistic situations." Aspects of rhetoric and rhetorical theory are based in interpretations of "problematic linguistic situations": rhetoric is designated here as the production of an interpretation; rhetorical theory is designated here as the analysis of the ethics, consideration of approaches to the situation, and probable outcome from different approaches. Rhetoric, however, continues long after we interpret the situation; that is, there is a process which may begin in an interpretation, but continues in the inventing, arrangement, styling, and delivering of a response. Furthermore, I suspect one can make too easy an argument for rhetorical situations as encompassing more than linguistic situations, starting with the example of the *Titanic* and working up to armed robberies and bad dates.

Finally, I am unsure what the last sentence of the above quote means: "As theoretical practices, hermeneutics involves placing a text in a meaningful context, and rhetoric requires the contextualization of a text's effects" (379). "The placing a text in meaningful context" may be hermeneutics as a variation of Gadamer's fusion of horizons, that is, the text is meaningful and understandable by an interpreter's historical context. But I am less sure of what is entailed when rhetoric needs "the contextualization of a text's effects." I believe, because of Mailloux's affection for rhetorical hermeneutics and the parallelism of his argument, that this is an argument for historical readings of rhetoric, for historically based rhetorical criticism as the basis of rhetorical theory. But this is only one model of rhetoric. Aristotle, for example, did just fine without an explicitly historical approach—in fact, he set up a two-thousand-year-old tradition from some second-rate psychology and close analysis of texts (tropes, syllogism, and topics). And rhetoric, as opposed to rhetorical theory, has to do with practices, not just readings of a text's effects.

Mailloux's collapsing of rhetoric and hermeneutics has implications for much of rhetorical theory—its canons, methods, theories of agency, institutional housing, writing and speaking pedagogy, and so on. Unlike Gadamer, who posits rhetoric and hermeneutics as connected by their focus on language, but still separated by distinctions, Mailloux offers a more passive rhetoric than we should accept. He offers a rhetoric in which one does not speak, but is spoken.

Rhetorical Invention, Hermeneutical Invention, Rhetorical Reading

It is my belief that we learn more about the concepts and our practices from the difficult task of differentiating hermeneutics and rhetoric than from that of collapsing them. If we follow Gadamer in perceiving interpretation as a re-production with the emphasis on production, creation, and invention, and if we want to differentiate rhetoric from hermeneutics, it is potentially productive to start with the differences between rhetorical and hermeneutical invention. I suspect there are many useful places to explore their differences, but let's start at the rhetorical start (with the first of the five parts of rhetoric). What are the processes of inventing a text within rhetoric and hermeneutics? Given their long histories, rather than comparing and contrasting rhetoric and hermeneutics, I turn to a site in no way simpler, but certainly more accessible: our everyday language. As Mailloux's difficulties demonstrate, connecting rhetoric and hermeneutics is a complex task: the concepts are broad, and the language slips. The differences in invention, so very hard to describe and defend, are suggested within our grammar. We can get a sense of the difference by probing the limited use of "invent" within everyday language and how we would paraphrase it in application. Who would announce, "I have invented an understanding"? Is it odder than to say, "I've invented an interpretation"? What would that mean? Is it as odd to say "I've invented a speech" or "I've invented a poem" or "I've invented a solution" or "I've invented a computer that invents interpretations and speeches"? The last claim, scientific invention, is the ordinary claim within language, but not the ordinary act of inventing; for while we may rarely claim to create a new mechanism, we constantly invent—in many ways—discursive forms. When we hear "I've invented a computer," we know that the speaker claims to have produced something that did not exist before, usually a machine, but not always so. It is unusual, but understandable to say, "Late last night, in my laboratory, I invented a new polymer chemical" or "I'm going to patent the organism I invented." Newness and complexity are part and parcel with scientific invention; we know scientific invention by its requirement of "origins" and originality. Alternately, because inventing discourse—or a response to discourse—is not part of our ordinary language and doesn't have such a stake in originality, we are less sure what it means to invent discourse. Unlike scientific invention, it is a daily act, but we know it less well. Discursive "invention" is jargon, and like all jargon, it is useful at times and obfuscating at others.

Even so, one does have some idea of how to respond to the jargon with ordinary language. One might correct a non-native speaker who claimed to invent a speech or poem with the paraphrases "I made up a speech," "I wrote a poem." One would understand here that she formulated words and ideas in some new order, maybe she even created a new word in her poem or found a striking metaphor. We might accept her claim to have "invented" a new word ("I invented a word") because of the newness of the single word and its demonstrable origin in her poem, but her claim to have invented a metaphor would puzzle us more, in part, because the words in a metaphor are not original: they existed within us, within our vocabulary, and we do not know how to assess her claim against the requirement of origin or novelty. At this point, we might give up the requirement of origin or novelty, recognizing that in her claim of having invented speeches or poems, there is no necessary surprising novelty, only the novelty of every utterance in a differing context. We paraphrase for her that "the speech is made up," "the poem written." The words themselves have no demonstrable origin (the OED aside). Inventing a speech might easily in this sense include responding to another's speech with the invention of a counterexample, another story in support or opposition, or a discussion of its logical fallacy, but that would not make us place her claim of invention in a different category than if she began with her own facts, story, or logic.

The non-native speaker's inventing of these texts can be explained by rhetorical invention, whether by the type found in Aristotle's *Topics* or *Rhetoric* or in a freshman handbook. Her claim doesn't suggest much about her theory of rhetorical invention. Hers might be a sophistic invention drawn from a moment in the flux between *dissoi logoi*. Or with equal plausibility, it might be an Aristotelian act of invention, where her ideas were shared with a community and simply recalled and arranged to effect the community's judgments. Or it might be one explained by any number of other theories; theories from Cicero to cognitive science to Barthes could be applied. The significant finding is this: "to invent a speech" is to make it up in some way that does not explicitly duplicate other speeches, and inventing speeches is explainable in existing theories of rhetorical invention. We do not need to decide which theory to recognize it as rhetorical invention.

What is meant in the cases of "inventing understanding" or "inventing interpretations" is far less clear because they move us further away from what invention means in both ordinary language and traditional rhetorical jargon. In the case of interpretation, to correct the non-native speaker and say "I made up an interpretation," "I wrote an interpretation" is not to claim an original form for one's words, but rather to acknowledge one's words as responding to a *particular* text, or set of texts. One might well ask her, "What did you interpret?"; and she might reply, "I interpreted a line from *Hamlet,*" or "I did a feminist reading of Shakespeare's tragedies." To say "I invented an interpretation" is, in effect, to claim a paraphrase or (and here's the rub) a reading of an extant text.

To claim a paraphrase is to claim hermeneutical invention, though a small and questionable one. Still, to paraphrase honestly is to put one's prejudices at risk. To claim a reading, however, is to claim an action that exists on a continuum between hermeneutics and rhetorical invention or at a site of their interactivity. If the non-native speaker has fused her horizon with that of the text, she has acted as a hermeneut. If she has responded to a particular text, but is claiming to have formulated words and ideas in some new order, unrecognized before (at least, by her), she may be, as a resisting reader or a critic, moving out of the hermeneutic realm and into the rhetorical. Maybe she has found an extension of metaphor or a contradiction in the text that lets her voice a new insight persuasively, maybe she has only extended the metaphor as the author intended, or maybe she has simply put her prejudices at risk and moved toward the text in a way she couldn't have before the hermeneutical moment. In this intersection between our everyday language and professional jargon, hermeneutical invention is a mediation which produces something, but the emphasis is on mediation, not production. The non-native speaker's intentions are necessarily part of, but subordinate to those in the text. She works well from a dictionary and a grammar book, and according to most hermeneuts, she probably can go so far as exegesis or explanation without entertaining novelty and rhetoric. The hermeneutically inventing non-native speaker risks her old prejudices to the horizon of the text: her meanings and the meanings of the text shift and fuse.

The non-native speaker who reads critically, suspicious of the text, full of her own intentions, reads with resistance and produces a meaning that allows her to take action. The invention in a rhetorical reading also is a mediation which produces something, but the emphasis is on production, not mediation. As resisting readers, an audience has a relationship to invention more like that of a rhetor than a hermeneut. The non-native speaker's intentions are foregrounded as prejudices to play out; what was a concern with the past, tradition, and prejudgments becomes a concern with intentions and future actions. In the tradition of feminist readings of the canon, interpretation can be rhetorical and inventive.

Rhetorical Reading and Rhetorical Invention

There is a gray area in interpretative invention where readers use their prejudices; here the border between hermeneutical invention (with risked prejudices) and rhetorical invention blurs. In *Inventions of Reading: Rhetoric and the Literary Imagination,* Clayton Koelb explicates this mode of reading, what he calls "rhetorical construction," and what others might characterize as "an aggressive form of interpretation imposing on innocent texts a reading that is 'invented' in the sense of being made up out of the critic's head with no thought given to what the texts themselves intend" (ix). Koelb connects inventive rhetorical reading with construction and use (invention and action). The nature of textual production from reading, though not the effect of the production, defines the act as rhetorical. Koelb sees the possibility of readings as infinite, but rhetorical readings as defined within a rhetorical tradition of construction and use, invention and action, deliberation and difference, not in the tradition of resolution or fusion of positions (244). Reading is invention, but not all reading is rhetorical reading that involves rhetorical invention.

If we return to the street corner where we left the by-now confused, non-native speaker and her need to "to invent an understanding," we leave discussions of rhetorical invention and enter the realm of hermeneutics, for inventing understanding is hermeneutical invention. In Gadamerian hermeneutics, the interpreter stops interrogating and manipulating the text and allows the text to interrogate our prejudices and intentions and finally to be applied in our present situation. For Gadamer, all *interpretation* is necessarily prejudiced, even the careful paraphrase, and there is no objective standard for an interpretation, not authorial intention, God's word, or scientific method. However, neither *interpretation* nor *understanding* is a matter of free play or subjective opinion. Understanding which includes interpretation also requires *application* or assimilation to our present, a fusion with our horizon. Understanding requires that the text becomes part of our being. He writes:

> Where there is understanding, there is not translation but speech. To understand a foreign language means that we do not need to translate it into our own. When we really master a language, then no translation is necessary—in fact, any translation seems impossible. Understanding how to speak is not yet of itself real understanding and does not involve an interpretive process. . . . Mastering the language is a necessary precondition for coming to an understanding in conversation. (*Truth and Method* 384–85)

And dialogue with a text, given common sharing and open questioning, is the basis of Gadamer's hermeneutics, a basis which requires that interpreters put at risk their prejudices. The point of hermeneutical invention is to produce a new position for the interpreter.

The invention of discourse then has many modes. I have outlined three, rhetorical invention, hermeneutical invention, and rhetorical reading. Invention within the act of interpretation is defined by two poles: by the interpreter's prejudices, privileged in rhetorical invention and diminished in hermeneutical invention, and by the interpreter's prejudices, risked in hermeneutical invention and accepted in rhetorical invention. A rhetorical theory of invention, one which acknowledges production and public discourse, need not stand free of interpretive theory, but it will need to recognize the meanings invented by an audience, an audience unlikely to be committed to ontological hermeneutics.

Rhetoric's increasing affiliation with textual reception, specifically Gadamer's hermeneutics, while increasing concern with discourse and text, potentially *diminishes* many aspects of textual production and of rhetoric: the invention of a speech act, interactions and negotiations of meaning and intentions between rhetors and audiences, cultural processes of deliberation and the resulting productions, formal aspects of the text, the contingency and immediacy of the rhetorical situation, the instrumental and subversive qualities of discourse, and the (ethical and unethical) power dynamics in any rhetorical interaction. If rhetoric comes "from the forum, the law-court, and the pulpit," these aspects of rhetoric must be articulated.

Notes

1. In "The Materiality of Discourse as Oxymoron: A Challenge to Critical Rhetoric," Dana L. Cloud provides a good background to and critique of rhetoric's increasing blindness to anything but the text.

2. In graduate school we talked about various unions: rhetoric and philosophy, rhetoric and feminism, rhetoric and linguistics. Inherent in these marriages is a belief that rhetoric was not substantial enough to stay single.

3. In a not unrelated protest, E. D. Hirsch writes that hermeneutics is used as "a rather vague, magical talisman" (*Aims* 19).

4. The literature in this area is large. The current discussion begins in a 1979 article by Michael Hyde and Craig Smith, is elaborated in the historical work of Rita Copeland and Katherine Eden, and explodes in the mid 1990s. Walter Jost and Michael Hyde's collection *Rhetoric and Hermeneutics* and George Pullman's special volume of *Studies in Literary Imagination* provide broad overviews of current questions and positions.

5. Gadamer writes, "[M]y real concern was and is philosophic: not what we do or what we ought to do, but what happens to us over and above our wanting and doing" (*Truth* xxvii).

6. In conversation Sue Wells made the excellent point that this has everything to do with the professional location of rhetorical work. Because "traditional rhetoric observations" can only be made in a social situation that offers all speech roles, potentially, to all speakers, there is little reason to be critical or analytic outside of production (and in a way, production is the best critique). But if the chance to "produce a desired effect" is asymmetrically distributed, then there is every reason for people to know how to analyze texts that they will never produce (for example, the Pentagon papers). The question for Wells, is how interpretative rhetoricians, such as Mailloux, do this in the absence of historicizing the *theory* (even if their interpretations are historically situated, their theoretical practices are not).

Works Cited

Cloud, Dana L. "The Materiality of Discourse as Oxymoron: A Challenge to Critical Rhetoric." *Western Journal of Communication* 58 (1994): 141–63.

Copeland, Rita. *Rhetoric, Hermeneutics, and Translation in the Middle Ages.* New York: Cambridge UP, 1991.

Eden, Kathy. "Hermeneutics and the Ancient Rhetorical Tradition." *Rhetorica* 5 (1987): 59–86.

———. "The Rhetorical Tradition and Augustinian Hermeneutics in *De Doctrina Christiana*." *Rhetorica* 8 (1990): 45–63.

Farrell, Thomas B. "From the Parthenon to the Bassinet: Death and Rebirth along the Epistemic Trail." *Quarterly Journal of Speech* 76 (1990): 78–84.

———. "On the Disappearance of the Rhetorical Aura." *Western Journal of Communication* 57 (1993): 147–58

Gadamer, Hans-Georg. *Philosophical Hermeneutics.* Trans. David E. Linge. Berkeley: U of California P, 1976.

———. "Rhetoric, Hermeneutics, and the Critique of Ideology: Metacritical Comments on Truth and Method." *The Hermeneutics Reader.* Ed. Kurt Mueller-Vollmer. New York: Continuum, 1992. 274–92.

———. *Reason in the Age of Science.* Trans. Frederick G. Lawrence. Cambridge: MIT P, 1981.

———. *Truth and Method.* 2nd rev. ed. Trans. Joel Weinsheimer and Donald G. Marshall. New York: Continuum, 1993.

Gaonkar, Dilip Parameshwar. "The Idea of Rhetoric in the Rhetoric of Science." *Southern Communication Journal* 58 (1993): 258–95.

Halloran, S. Michael. "On the End of Rhetoric, Classical and Modern." *College English* 36 (1975): 621–31.

Hirsch, E. D. *The Aims of Interpretation.* Chicago: U of Chicago P, 1976.

Hyde, Michael, and Craig R. Smith. "Hermeneutics and Rhetoric: A Seen But Unobserved Relationship." *Quarterly Journal of Speech* 65 (1979): 353–63.

Jost, Walter, and Michael Hyde. *Rhetoric and Hermeneutics in Our Time: A Reader.* New Haven: Yale UP. 1997.

Koelb, Clayton. *Inventions of Reading: Rhetoric and the Literary Imagination.* Ithaca: Cornell UP, 1988.

Leff, Michael. "The Idea of Rhetoric as Interpretive Practice: A Humanist's Response to Gaonkar." *Southern Communication Journal* 58 (1993): 296–300.

Mailloux, Steven. "Articulation and Understanding: The Pragmatic Intimacy Between Rhetoric and Hermeneutics." *Rhetoric and Hermeneutics in Our Time: A Reader.* Ed. Walter Jost and Michael Hyde. Princeton: Yale UP, 1997. 378–94.

———. *Rhetorical Power.* Ithaca: Cornell UP, 1989.

Palmer, Richard E. *Hermeneutics.* Evanston: Northwestern UP, 1969.

Pullman, George, ed. *Studies in the Literary Imagination* 28.2 (1995).

Invention and Inventiveness

A Postmodern Redaction

YAMENG LIU

The rise of postmodernism as a dominant intellectual trend since the 1970s and the prevailing of what may be called the postmodern ethos in our time have radically changed the context of contemporary rhetorical practices. As a result, the traditional terms in which we used to talk about invention have been rendered inadequate. In the light of a postmodern epistemology, for example, it is no longer tenable to regard an "argument" or a "means of persuasion" as something "out there" waiting to be "discovered," or to think of the rhetor or the audience as an autonomous agent existing prior to and independent of the rhetorical process. A postmodern sensitivity to the power/discourse nexus has alerted us to the fact that rhetorically significant "topics" or "commonplaces" have to be *authorized* by dominant political, institutional, or cultural formations in order to function as such. And the postmodern preoccupation with communication ethics has even given rise to the issue of whether "persuasion," traditionally pitted against coercion, is as free from force as it sounds. These new perspectives have been asserting themselves for some time as *the* problematics within which to foster a postmodern understanding of invention. Yet despite the significant impact they have been able to exert on contemporary thematizations of rhetoric, such an understanding is still far from establishing itself. As the currently "standard" definitions of invention show, our understanding of this central rhetorical concept remains trapped in a modernist, rather than postmodern, mode of thinking.

In scholarly works on classical rhetoric, it is conventional to set forth the meaning of invention in terms of "creation" and/or "discovery," the two words that together shaped the way discursive production was understood and described prior to the postmodern turn in contemporary discourse. It is not uncommon for *inventio* to be defined as "[that] part of rhetoric concerned with the *discovery* of arguments" (Conley 317, emphasis added) or as "[the] generation, or *creation,* of ideas for a given discourse" (Welch 169). The same is found in works on modern rhetorical theories. Gerard Hauser in his *Introduction to Rhetorical Theory* suggests that invention should be understood as "*discovering* what might be said on a subject to persuade an audience" (61, emphasis added), and he begins his chapter on rhetorical invention with an elaborate discussion of the central importance of being *creative* (57–61). And it is by no means unusual to find in a collection of contemporary psychological and philosophical perspectives on the subject of creativity an essay titled "The Possibilities of Invention," which frequently uses "discovery" as a variation for either "invention" or "creation" (Sternberg 362–85).

Such a practice is so entrenched that not only does it strike us as "natural," but many of us are unable to find any significant issue in it even when questions are raised about its underlying assumptions.[1] For many scholars, what is involved here is nothing more than a loose but largely innocuous use of three familiar terms that have acquired, through usage, similar, and at times overlapping, senses. The treatment of these terms as more or less synonymous dates back to classical antiquity, when rhetoric was still the dominant art. And there does not seem to be much ground for believing that the mutual substitution of the terministic trio is indicative of something as serious as a failure to think through what it means to invent, or as grave as a clinging to old ways of understanding how discourse is produced.

This perception is understandable, but hardly warranted. Semantically speaking, to "discover" is to make visible or known something that, though hidden and unknown previously, has always been "out there," something whose existence prior to its "discovery" is immediately recognized because of its fundamental fit with the order of things already familiar to us. To "create," on the other hand, suggests bringing into being something that has never before existed, some strange entity snatched *ex nihilo* which is, presumably, completely different from whatever has been accepted as part of the "reality," and which, therefore, refuses to conform to our habitual scheme of conceiving the world. In making a "discovery," we identify the unknown with the known. In making a "creation," we differ radically from the known. If we translate all these semantic distinctions into rhetorical terms, invention-cum-discovery would presuppose a mode of discursive production that is decidedly oriented toward finding new ways for reaffirming what both the speaker and the audience hold, consciously or unconsciously, as "true" or "valid," whereas invention-cum-creation would imply an orientation in just the opposite direction. To invent in this latter sense is to search for ways to define a fundamental difference from whatever has already been accepted as part of the structure of reality.

Contrary to our impression that invention has "always" been defined this way, classical authors in general did not associate it explicitly with either "discovery" or "creation," or with what these two words signify to us now. The concept of invention remains understood throughout Aristotle's *Rhetoric.*[2] Both Plato (236a) and Quintilian (3.3.1) make references to invention in their texts, but never bother to define it. The most explicit definition of the term is found in *Ad Herennium,* where it is said to be "the *devising* of matter, true or plausible, that would make the case convincing" (1.2.3, emphasis added).[3] The synonymization of the three terms in the defining of *inventio* was clearly a later development. While there is no denying the polysemous character and semantic overlap of the three words in their common usage, it is equally true that "discovery" and "creation" had been appropriated by theorists of discourse since the sixteenth century, purified of their fuzzy, everyday significance, and enshrined as the god-terms for the empirico-objectivist model of composing and its alter ego, the romanticist model of discursive production.

At least since Bacon, a self-consciously defined notion of "discovery" has been granted centrality in conceptions of writing. Bacon, as Lisa Jardine in her study of his art of discourse observes, is "bound to consider . . . dialectic and rhetoric as second-class studies, because he is so deeply preoccupied with *discovery* as the primary mode of human experience" (170). It is precisely on the basis of a discovery-centered conception of writing that Bacon dismisses "rhetorical invention" as a misnomer in *The Advancement of Learning,* on the ground that "invention of speech or argument is not properly an invention, for to invent is to *discover* that [which] we know not, and not to recover or resummon that which we already know" (58, emphasis added). He later relents a little, observing condescendingly in the same text that because we call it a "chase of deer" whether the chasing takes place in an "enclosed park" or in "a forest at large," (and because rhetorical invention has "already obtained the name") we may as well "let it be called invention" (58). Yet it is clear that for Bacon and like-minded theorists, invention should be properly understood as the "discovery" of the "deer," that is, the discovery of something hidden, invisible, elusive yet always already existing out there, "treasured up in the bosom of nature" (129). As such, rhetorical "invention" can and should only be understood as a shadow of the "real" invention because it at best "recovers" something already existent within a known enclosure and does not lead to any genuine "discovery."

At least since Coleridge, a self-consciously radicalized notion of "creation" has acquired a comparable privileged status in the conceptualization of writing. Coleridge recognizes what he terms "the whole tremendous difficulty of a Creation *ex nihilo* [out of nothing]," yet insists that the concept cannot refer to anything else, for, in his words, if it is "*ex aliquo* [out of something], how could it be Creation?—and not in all propriety of language [be] Formation or Construction?" (572). For him, to "create" is necessarily to "will causatively" something entirely new, entirely different, without prior existence, into being (573). Anything short of this belongs to acutely different categories such as "formation" or "construction." These latter categories are definable only in terms of their failure to achieve what "creation" alone is able to accomplish. And unlike Bacon, Coleridge did not even bother to mention rhetorical invention, which is suggestive of the extent to which rhetoric and the concept of invention had been decentered by his time.

Coleridge's subjectivist project of conjuring up something out of nothing would seem to be poles apart from Bacon's objectivist quest for revealing a hidden or delayed presence in nature. Yet for all their apparent epistemological differences, they have much in common. In privileging the genuine "invention of sciences" over the supposedly counterfeit "invention of speech and argument," or in promoting the *bona fide* "creation" over the supposedly diluted "construction," they both affirm, as *the* approach for generating new discourse, a single-minded search for the unknown, the different, the novel. Moreover, they both set as the object of this search something which is unambiguously authentic or true, whether this authenticity or truth is certified by a collective, even universal, recognition, by the immediacy of the creative act, or by the authority of the creator. Both models share, therefore, what Stephen Toulmin identifies as the spirit of the modernist mode of discourse, that is, the belief in the possibility and the necessity of "a clean slate and a fresh start" for our discursive practices (177).

Whether we like it or not, this belief has been inscribed in the twin concepts of "discovery" and "creation," and any uncritical use of these terms carries with it these modernist connotations. Since to link invention with these two terms is to see the former from the perspective of the latter, the linkage is never innocent of conceptual and theoretical implications for rhetorical practices. Bacon and Coleridge themselves, in fact, would have agreed fully with this view. Their discussions display such a sensitivity to semantic differences among apparently synonymous key words that, for them, the choice of vocabulary was by no means a small matter. Bacon would countenance a continued use of "invention" in reference to writing only on condition that two fundamentally different *kinds* of invention be clearly differentiated. For Coleridge, there was an essential difference between the term "creation" and words such as "formation" or "construction," and the treatment of these terms as near synonyms was never sanctioned.

Definitions of "pivotal terms," as I. A. Richards tells us, are "key positions" in any theoretical formation (41). The way terms are selected, redefined, and rearranged on a scale of value often constitutes the ground on which the most decisive battle between competing theories or paradigms is fought. Therefore, one who wishes to promote a new theoretical model or to challenge an old one cannot afford to be careless about the terms he or she is using. This was true of the originators of the modernist discourse theories. It is true also of the pioneering figures in the postmodern discourse. Talking about the "stakes" involved in his project of subjecting familiar philosophical languages to a close reexamination, Jean-François Lyotard points out that "thought, cognition, ethics, politics, history or being, depending on the case, are in play *when one phrase is linked onto another*" (xii–xiii, emphasis added). In stressing the interrelatedness between "thought," "ethics," or "politics" and the way "phrases" are "linked," Lyotard merely reaffirms the conclusion which Raymond Williams reached in his earlier study of culture and society.

Williams finds that major social and cultural changes are always correlated with major changes in the language people speak. "When we come to say 'we just don't speak the same language,'" according to Williams; we actually mean that "we have different immediate values or different kinds of valuation, or that we are aware, often intangibly, of different formations and distributions of energy and interest" (11). Such transformation of "values," "valuation," or "interest" finds a focused expression in the way we select and use what Williams calls "keywords," those central terms which are either "significant, binding words in certain activities and their interpretation" or "significant, indicative words in certain forms of thought" (15). The understanding, consolidation, or promotion of new experiences, activities, or practices thus becomes for Williams very much a matter of setting down "available and developing meanings of known words" and of sorting out "the explicit but as often implicit connections which people were making, in . . . particular formations of meaning—ways not only of discussing but at another level of seeing many of our central experiences" (15).

This is precisely what Derrida tries to do when he takes notice of "a revival, a new fashionableness, and a new way of life," experienced in our time by "that tired, worn-out, classical" term "invention," and wonders out loud why this old rhetorical concept should "[impose] itself . . . more quickly and more often than other neighboring words" such as "'discover,' 'create,' 'imagine,' 'produce,' and so on" (*Derrida Reader* 217–18). Derrida discriminates among these words in the context of a reflection on Paul de Man's deconstructive rhetorical theory, and his paradigm case for invention is the musical genre bearing that name rather than the "devising" of "matter," capable of persuading an audience in classical rhetoric. Even though his interest lies not so much in the topic of invention itself as in exploring new ways of describing and justifying his "movement of deconstruction" (*Derrida Reader* 217),[4] Derrida's study nevertheless points to the crucial difference the choice of one key word over another would make. In particular, he turns our attention to the terministic trio in question and stresses their conceptual incompatibilities when he asks rhetorically, "Why is it that *invention* cannot be reduced to the discovery, the revelation, or the unveiling of truth," any more than "it can be reduced to the creation, the imagination, or the production of the thing?" (*Derrida Reader* 200).

By associating "discovery" with the "unveiling of truth" and "creation" with the "[fictional or fabulous] imagination . . . of the thing," and by insisting on the irreducibility of invention to either of these two concepts, Derrida undercuts the common practice of defining invention *in terms of* discovery or creation, thus deconstructing the frames of reference which Bacon, Coleridge and other modernist theorists of discourse had established. Derrida insists that invention is irreducible to either "discovery," as the "unveiling of truth," or "creation" as the "[fictional or fabulous] imagination of a thing." Their theoretical models are no longer deemed reputable by most theorists of discourse. Yet if the survival and persistence of the key words associated with their models is any indication, at a deeper level, people continue to make what Williams terms the "intangible" or "implicit" connections in the "particular formations of meaning" which they helped to set up. This is why "[within] an area of discourse that has been fairly well stabilized since the end of the seventeenth century in Europe," as Derrida observes,

> there are only two major types of *authorized* examples for invention. On the one hand, people invent *stories* (fictional or fabulous), and on the other hand they invent *machines*, technical devices or mechanisms, in the broadest sense of the word. Someone may invent by fabulation, by producing narratives to which there is no corresponding reality outside the narrative, . . . or else one may invent by producing a new operational possibility . . . [So that we have] only two possible, and rigorously specific, registers of all invention today. . . . Whatever else may resemble invention will not be recognized as such. (*Derrida Reader* 204)

The fables and machines to which Derrida refers are emblematic of the two modes of discourse production signified by "creation" and "discovery." The desynonymization of the terministic trio of invention, creation, and discovery enables the kind of theoretical reflexivity necessary for resolving many long-standing issues in rhetoric and composition. With a heightened awareness that to invent, contrary to what Bacon teaches, is *not* merely to "discover," we would be more wary of the folly of presupposing a "deer" or some other sort of prior existence independent of the will or intent of the "discoverer" when we talk about invention. We would be in a better position to detect the assumption of this "deer" in whatever form it happens to assume, whether it be a "rhetorical situation" understood as a *given,* the "audience" construed as the "real" individuals out there, the "discourse community" defined as a stable, easily extractable set of interpretive strategies or conventions, or even the "power" to be sought, in Foucault's words, "in the primary existence of a central point, in a unique source of sovereignty from which secondary and descendent forms would emanate" (93).

Similarly, a heightened awareness that to invent (contrary to what Coleridge suggests we should all strive for) is *not* to "create" would make a difference in both our conceptions and practices of rhetoric. Such an awareness would suggest, for example, that we think twice before we focus our efforts only on the cultivation of "real insights," "original perspectives," "unconventional thinking," transgressive acts, or authentic "inner voices"—as if it were possible in the act of invention either to effect a clean break with what is conventional, ordinary, commonsensical, and traditional or to go beyond the "text" in such a way as to transcend our socio-ideological situatedness and cultural inheritances. The notion of originality, as James C. McKusick points out in his critique of Coleridge's preoccupation with linguistic origins, is ambiguous and paradoxical: "its primary meaning is the ability to originate, to create *ex nihilo;* but it also hints at a return to origins, and thus a reliance on primary source materials." Moreover, the related emphasis on "the autonomy of individual creation" has also created an "uneasy tension" with the knowledge that "any creative activity must emerge from a larger tradition from which it derives the greater part of its materials and insights" (53). In other words, what is "new" is always already saturated with "traces" of the old, what is "unique" saturated with "traces" of the common, what is "different" saturated with traces of the same. Indeed, one wonders if we as rhetoricians should continue to encourage writers to become "creative," with all its established connotations of radical difference and newness.

While it makes little theoretical sense to continue to privilege the "original" *at the expense of* the common or the usual in our discursive or pedagogical practices, creativeness as an ideal has been so solidly institutionalized that to stop preaching it to our students is bound to cause a serious disruption to the current system of rhetorical values. Yet to "have different immediate values or different kinds of valuation," as Williams tells us, is precisely what we should do in order to speak a different language (11). It is theoretically imperative to decenter the two "rigorously specific registers of invention" to which Derrida refers and to promote "inventiveness" in place of "creativeness." To be inventive is to strive for the new without attempting a clean severance with the old and to search for the unique through an identification with the common; it is to try to achieve originality, with the understanding that the more original a perspective is, the more deeply it is rooted in the conventional. To promote inventiveness is to insist that it is as important to "recover or resummon that which we already know" as it is to "discover that [which] we know not." There is not, nor can there be, any fundamental difference between "creation" and "formation" or "construction."

With inventiveness as a rhetorical value, rhetoric no longer has to feel apologetic for failing to measure up to the standard of originality that is central to the concept of discovery or creation. Instead, creation and discovery may be seen as two analytical aspects of invention, definable only in reference to the latter. In other words, rather than conceiving of invention as a defective, diluted, version of either creation or discovery, creation and discovery each signify only one of the two mechanisms indispensable to discourse production. The promotion of "inventiveness" would thus make a decisive contribution to the reversal of the traditional hierarchy among the three key terms, making it possible for us to articulate a postmodern model of composing.[5]

Notes

1. Derrida is sensitive to the subtle yet important difference among "invent," "discover" and "create," and as Young and Liu point out, this terministic trio actually signifies "three quite different orientations in understanding discursive production." See Young and Liu xii–xvi.

2. The meaning of invention is clearly implied in Aristotle's introductory remarks as well as in his definition of rhetoric in *Rhetoric* 1355b. In all these places, he uses *theoresai,* "to be an observer of and to grasp the meaning or utility of," to refer to the act of inventing. The Greek verb has been turned into the English verb "to see" by Kennedy in his new translation of the Aristotelian text. See Kennedy 35–37. Aristotle, of course, talks about invention—often indirectly—in many other ways.

3. The "devising" in this definition translates *excogitatio* in the original text, which is cognate to *excogito,* "to scheme, devise, contrive."

4. While it seems obvious that the Derridean concept of "deconstruction" is antithetical to the concept of composition and hence incompatible with that of invention, a more sympathetic reading of Derrida's conception of invention is possible. Derrida does not believe in the possibility of "stepping outside" the "text" (i.e., the horizon of signification of any discourse), and of creating something entirely new. He readily admits the crucial role a central core of any signifying structure plays in the "[orienting], [balancing] and organizing" of the system that "permit the play of its elements inside the total form." Yet he is also keenly aware of the tendency of this central core to "[close] off the play which it opens up and makes possible" (*Writing and Difference* 279). His emphasis on the principle of "undecidability" and on the mode of "deconstruction" can thus be seen as a call for preventing the hardening of this core into a "given," which inevitably leads to the "closing off" of inventive possibilities. Such a reading is consistent with his fascination with the classical term "inventio."

5. In presuming a double-gesture of trying to identify and differ simultaneously, "inventiveness," when established as the central value of a postmodern rhetoric, may also offer a solution to another axiological problem. This is the theoretical and pedagogical difficulty caused by the separate and largely unexamined valorization of reflexivity and criticality as the twin discursive virtues of our time.

As they have been understood, the two are hardly compatible to each other. Whereas reflexivity dictates, as the way to invent, that the inventor turn inward and be concerned primarily with self-probing and self-positioning, criticality directs attention outward and upward. In itself, neither of these terms fits in with what has generally been recognized as the postmodern thinking. To be "reflexive" in actual textual practices frequently amounts to admitting that the inventor is caught and trapped in a web of power relations, with little hope of opening up new room for the play among these relations. And what is known as the reflexive mode often makes the problematic presupposition that deep-seated motivating interests are easily accessible to the inventor herself (see Liu, "Rhetoric and Reflexivity"). Similarly, an uncritical promotion of criticality often leads to the facile assumption that a truly oppositional position outside whatever dominant formation happens to be involved, from which one can attack the system with great ease, is readily available to the critic.

Works Cited

Aristotle. *On Rhetoric.* Trans. George A. Kennedy. New York: Oxford UP, 1991.

Bacon, Francis. *Advancement of Learning, Novum Organum, New Atlantis.* Chicago: Encyclopedia Britannica, 1952.

Coleridge, Samuel Taylor. *The Oxford Authors: Samuel Taylor Coleridge.* Ed. H. J. Jackson. Oxford: Oxford UP, 1985.

Conley, Thomas M. *Rhetoric in the European Tradition.* Chicago: U of Chicago P, 1990.

Derrida, Jacques. *Writing and Difference.* Chicago: U of Chicago P, 1978.

———. *A Derrida Reader: Between the Blinds.* Ed. Peggy Kamuf. New York: Columbia UP, 1991.

Foucault, Michel. *The History of Sexuality.* Vol. 1. New York: Vintage, 1980.

Hauser, Gerard A. *Introduction to Rhetorical Theory.* New York: Harper, 1986.

Jardine, Lisa. *Francis Bacon: Discovery and the Art of Discourse.* Cambridge: Cambridge UP, 1974.

Liu, Yameng. "Rhetoric and Reflexivity." *Philosophy and Rhetoric* 28 (1995): 333–49.

Lyotard, Jean-François. *The Differend: Phrases in Dispute.* Minneapolis: U of Minneapolis P, 1988.

McKusick, James C. "'Singing of Mount Abora': Coleridge's Quest for Linguistic Origins." *Critical Essays on Samuel Taylor Coleridge.* Ed. Leonard Orr. New York: Hall, 1994.

Richards, I. A. *Richards on Rhetoric.* New York: Oxford UP, 1991.

Sternberg Robert J., ed. *The Nature of Creativity: Contemporary Psychological Perspectives.* Cambridge: Cambridge UP, 1988.

Toulmin, Stephen. *Cosmopolis: The Hidden Agenda of Modernity.* New York: Free Press–Macmillan, 1990.

Welch, Kathleen E. *The Contemporary Reception of Classical Rhetoric: Appropriations of Ancient Discourse.* Hillsdale: Erlbaum, 1990.

Williams, Raymond. *Keywords: A Vocabulary of Culture and Society.* Revised ed. New York: Oxford UP, 1985.

Young, Richard E., and Yameng Liu. Introduction. *Landmark Essays on Rhetorical Invention in Writing.* Ed. Richard E. Young and Yameng Liu. Davis: Hermagoras, 1994. xi–xxiii.

Institutional Invention

(How) Is It Possible?

LOUISE WETHERBEE PHELPS

In this chapter I want to explore several broad questions with respect to higher education: Is institutional invention possible? What are the conditions that enable it, and how can they be created and sustained? What are the obstacles to institutional invention? How can academic leadership foster institutional invention?

Institutional invention has two complementary interpretations in my inquiry. First, it applies reflexively to the academic institution itself: forming and reforming its ideals, governance structure, financial resources, curriculum, and so on. For a college or university, invention in this sense might mean designing new general education requirements, "restructuring" its mission and budget, or developing a student advising system. Second, it refers to the idea that an educational institution can deliberately organize itself as a hospitable environment for people to engage in their own creative work and learning. In either case, innovation might involve either a whole institution or one of its parts or functions. I am interested in the relationships of complementarity, interdependence, and possible conflict between these two aspects of institutional invention as they come together in the academy today. My topic also entails the problem of leadership: What part do leaders play, and how is leadership to be understood, in relation to institutional invention? I raise this question with a special concern for the dangers inherent in faculty disengagement from leadership roles and leaders during a period of intensive change for American colleges and universities.

Whence come such questions? The need for institutional invention first became visible to me when I was hired to lead composition faculty in building a new university writing program at Syracuse University. From the start I perceived this collective enterprise and my own leadership role in analogies drawn from rhetoric and writing—composing, communication, the creation of genres, collaborative inquiry, invention. In curriculum and pedagogy, it was really a task of deep revision, since writing had been taught in a particular paradigm at the university for over fifteen years and would continue to be taught by many of the same people who would ultimately invent the new curriculum. Organizationally, it meant designing from scratch every feature and element of a decidedly nontraditional academic unit from its budget, space, and staffing to its mission, communications, and social architecture.

After six years as director, I stepped down to become a faculty member in the writing program we had successfully built into a department. This post-directorship offered an extraordinary opportunity to observe the unexpected consequences of institutional invention, to reexamine my own actions and assumptions as a change agent, and to develop a more complex understanding of innovation and change processes. As the program faced new, more stringent circumstances, it thrived in ways that both exceeded and fell short of our earlier visions. My attention shifted from what came to seem relatively easy—introducing new ideas and structures in a time of excitement and expansion—to the difficult, frustrating labor of consolidating and institutionalizing those changes. Later, realizing that inscribing a revolution ultimately recreates stasis, I changed focus again: from preserving the innovations per se to sustaining a climate of invention—an environment that would encourage and support creative work and learning by everyone in an ongoing way. Reframing the task this way transforms and complicates assumptions about leadership.

As I learned of the high stakes for constructive change in higher education, I redefined, expanded, and recontextualized my questions about institutional invention from writing programs to the broader arena of educational reform. What I bring to the following discussion—and what I believe the field of composition and rhetoric can offer higher education—is this dual perspective on institutional change: a writing program administrator's practical experience in educational reform and innovation combined with a rhetorician's theoretical perspectives and frameworks for understanding invention.

The Need for Institutional Invention in Higher Education

At the beginning of the twenty-first century, we find ourselves in the midst of accelerating cultural changes that demand constant innovation and adaptation to new challenges. Core faculty in higher education, cushioned by tenure and the tremendous conservatism of the academy, have felt these pressures later and less dramatically than professionals in sectors such as health care or business. Change factors affect institutions unevenly, having their mildest and most delayed impact on those of greatest prestige and wealth. Thus, while academics in some schools, regions, or disciplines have suffered keenly from deteriorating conditions and attitudes toward higher education, many others were long able to ignore such forces or treat them as only temporary or localized.

They had good reason for their faith in the immutability of the academy. Its fundamental structure and values hadn't changed much in the United States since the late nineteenth century and, in some respects, since the earliest foundings of universities. (Twenty years ago, Clark Kerr observed that of sixty-six institutions existing in 1530 and still extant, sixty-two were universities [qtd. in Zemsky and Massy 41].) Yet this continuity and stability had allowed the American higher education system to develop unparalleled diversity and (paradoxically) supported unfettered innovation and perpetual change in the knowledge disciplines and curricula it housed. Now, however, many observers suggest that such constancy is no longer adaptive, but is making the academy inflexible in meeting a situation of great fluidity and new societal demands. Historians point out, too, that the research university that dominated U.S. education during the twentieth century was itself an innovation responding to social pressures, notably the late-nineteenth-century need to pursue science as a means of creating advanced knowledge rather than as a practical, empirical art (Kevles). The ability to remake itself radically in times of social transformation is itself part of the tradition of the American academy (Rudolph).

Since the early nineties, prophetic leaders and scholars of higher education have been analyzing the multiple, synergistic forces affecting education, warning of their profound, far-reaching consequences and trying to stimulate proactive reform by individual institutions and by leaders of the collective enterprise.[1] Their consensus is that this is not a passing phenomenon to be waited out or fixed by tinkering, but a constitutional crisis that challenges American higher education once more to redefine its purpose and its social contract with its many publics. In the familiar cliché, higher education, like every other sector of society from health care and business to government, needs to be "reinvented" if it is to survive and thrive in this new century, when it will no longer hold a monopoly on the production and dissemination of knowledge.

Most academics wince at "reinvention" as a corporate buzzword and (with some justification) fear the specter of business taking over the academy and corrupting its core values and mission. But for a writing teacher and administrator, the idea of "reinventing" the academy evokes a different frame of reference—that of rhetoric. To what extent might academic leaders or collectives be thought of as "composing" or "revising" an institution in response to an exigence, in situations defined as rhetorical by their uncertainty, indeterminacy, probable reasoning, and conflicts of value? One might even suspect that a great deal of the work of an academic leader is rhetorical in the stricter sense: discursive, accomplished through spoken and textual dynamics.

But the diminished notion of agency developed in composition and rhetoric within the last ten years provides weak support for theorizing such a role for faculty leaders as composing or reinventing our institutions. (Similarly, as we will see, traditional academic culture offers little help for such a concept in its understandings of faculty activity, faculty role, governance, and relationships between faculty and administration.) Typically, in recent composition studies, agency within institutions (or cultures) is imagined primarily in terms of conflict or resistance. Institutions, and leadership of or within them, are construed primarily as instruments or channels of (coercive) cultural power rather than of constructive action or invention, while creativity is understood as a subversive or countercultural force (in effect, the individual academic or student against the institution) rather than as a potential feature of the institution.

There is historical truth in this view with respect to academe as we have known it in the last fifty years, at least in the ideal that has been transmitted over many academic generations. (That is one reason that idealized notions of collaboration and consensus, proposed by some feminists in rhetoric and composition as principles for academic leadership, remain utopian.) "Inventiveness" or creative force within the academy has been thought of as a feature or quality of the autonomous faculty—either of individuals or of research groups; and it has been powerfully attributed to and located not within the local institution per se, but within the transinstitutional disciplinary domain, where it applies only to scholarly research. Higher education institutions themselves, as noted, are still deeply conservative in structure and concept. So, within the long-standing and still current paradigm of higher education, the idea of "institutional invention" (that is, of institutions being holistically inventive or continually self-inventing) seems to be an oxymoron.

Yet present circumstances demand that institutions themselves must become radically inventive, in these two distinct but interdependent senses:

- to thoughtfully—not reactively—transform themselves (their goals, organization, financing, relations to their publics, and so on) in order to meet new social expectations and needs without losing the qualities they value most
- to enable continual innovation and adaptation in any domain by those populating or served by the institution: not just by faculty, but by students, staff, administrators, and institutional partners in the community; not just in research but in all possible academic roles and services.

Insofar as local institutions, or units and domains within them, succeed in becoming inventive in these terms, they may contribute to the work of reforming higher education itself as a system, an institution in the more abstract sense.

Considerable work, both conceptual and practical, has been done by policy makers and administrators toward this end, but it has not yet become truly a faculty enterprise. Faculties have responded variably to the heralds of change and to the would-be reformers: with indifference, cynicism, or denial; with grief, rage, and resistance; rarely, with a sense of control, confidence, or shared personal investment in local changes. Only recently have faculties, refreshed by increasing numbers of a new generation, begun genuinely to believe (even if many deplore) that the academy is subject to the same powerful forces that are restructuring other sectors of society. Presented with unmistakable evidence that change is inevitable and irreversible, faced with the ambiguities of reform and the dangers of inaction, professors are beginning to wonder what responsibility they need take for shaping such a transformation of the academy, with all its destructive and creative potential.

The challenges facing higher education call for creative faculty leadership in and of institutions. We are all simultaneously actors and acted upon within an intensely interactive (though sometimes invisible) network of events and forces. Those of us working in the academy must confront questions regarding our own responsibility and effectivity as agents of institutional change or, perhaps, as guardians of academic tradition—or both at once. As faculty careers progress we find ourselves playing such institutionally empowered roles as project coordinators, curriculum designers, program directors, department chairs, committee chairs, deans, or central administrators. Some of us work in venues outside the institution proper, writing and speaking as public intellectuals or taking positions in higher education organizations. Even as ordinary academic citizens we must help select, support or resist, judge, even cultivate and develop leaders within our institutions. All these activities fall within the broader understandings of "faculty role" and suggest that leadership contributions may be made at all levels of rank and position, often from those who hold no formal administrative position.

For professionals in the field of composition and rhetoric, institutional invention in the context of change in higher education presents a particularly salient and congenial set of problems. Salient because composition specialists face the challenges of administrative leadership earlier and more often than faculty in most disciplines; their professional expertise and hands-on experience increasingly include cross-institutional knowledges and responsibilities for innovation far broader than writing itself (for instance, service learning, multimedia technology, and interdisciplinary teaching). Congenial because a rhetorician's training and concerns with inquiry, invention, and problem solving in writing are adaptable to analyzing what I am calling "institutional invention." I will work here at the intersection of these two forms of knowledge: a rhetorically informed understanding of invention and a practical experience of leadership as a site for creative action.

Concepts of Invention

At first sight it may appear that my questions and concerns are far removed from a traditional understanding of rhetorical invention. However, in the revival of invention within contemporary composition and rhetoric there has always been a systematic ambiguity in the concept between a generic notion of creativity or discovery (as in science and the fine arts) and a highly specific art of finding proofs for spoken or written arguments. The distinction isn't sharp; and definitions and uses of invention in rhetoric and composition tend to waffle between the two senses or fall somewhere in the continuum between the two poles.[2] Typically, studies of rhetorical invention (e.g., Young, Becker, and Pike) that tilt toward the general end emphasize creativity as an inquiry phase in a process of making new knowledge or meaning to which discourse becomes instrumental. These theories associate invention with indeterminate situations, problem formulation and problem solving, and a process of inquiry, especially as applied in knowledge disciplines. They hold what Kaufer and Butler call the modern "novelty standard" for invention in contrast to the classical criterion of cultural resonance: "on the novelty standard, persuasion means little [in evaluating the quality of an idea] if it is not also seen as innovation, as ideas pushing the boundaries beyond what is currently known or archived" (3). The narrower or more classical definitions (e.g., Corbett) stress the composition of specifically discursive ideas appropriate to, organized by, and directed toward written or spoken genres and rhetorical situations specifying purpose, audience, exigence, and so on.

Karen LeFevre in her study of invention as a social act adopts a similar distinction between a "broad" and a "narrow" view of rhetorical invention. She chooses the broader interpretation because "[c]onceiving of rhetorical invention as a search for wisdom—a search that involves analyzing subjects, audiences, and problems as well as generating and judging ideas, information, propositions, and lines of reasoning—aligns rhetorical invention closely with inquiry or with 'invention' in the generic sense: the creation of what is new in any discipline or endeavor. Rhetorical invention becomes a species of invention in general" (2). Noting that she could either use rhetoric as a lens for understanding invention in any endeavor or take the converse approach, LeFevre decides to study invention in the generic sense for the insights it will yield into rhetorical invention.

Although, like hers, my approach is steeped in rhetorical ideas and I am mindful of the metaphorical resonances between rhetorical and institutional invention, I don't intend to use rhetorical theories directly to derive a concept of institutional invention. One reason is that, with a few exceptions, they don't seem readily adaptable or extendable to understanding invention as an attribute of a system. Instead I will follow LeFevre's strategy by exploring institutional invention as a species of invention in general. From this perspective, rhetorical and institutional invention are parallel or analogous arts: each is a specialized type of invention with its own sites, materials, audience, products, and so on. Other relations between them are possible, as well. For example, institutional invention must be a highly rhetorical phenomenon itself, insofar as organizations are internally networked and their inventiveness depends on communication. I have already suggested that rhetorical invention provided me with powerful metaphors to understand program building and its constituent activities as institutionally inventive. Only time can tell whether a more developed art of institutional invention may in turn illuminate the concept and practice of rhetorical invention, perhaps in its collaborative or collective aspects. Certainly it should help us to understand the design of programs and curriculum, administration of writing programs, classroom teaching, and student learning as domains of creative activity.

These observations frame an inquiry that will proceed in several steps. First, I will develop a concept of invention as an emergent phenomenon of complex systems. Second, I will ask how this concept might change our perspective on leaders' responsibilities and generate new questions about leadership. Third, I will examine the barriers to institutional invention in current academic culture. My goal is to set the stage for developing a practical art of institutional invention, with special attention to the role of leaders at all levels and degrees of power and authority in fostering it. I hope this preliminary inquiry will contribute to that effort by formulating the problem in fresh terms and generating an array of productively specific questions.

Invention as a Phenomenon of Complex Systems

In order to understand how inventiveness (including self-invention) might characterize an organization, group, or institutional setting, we need a concept of creation or discovery as a holistic feature of cultural systems. At first sight, LeFevre's study of invention as a social act appears at least in principle to offer such an account. She distinguishes four interpretations of invention as social, on a spectrum from the individual to the collective: the Platonic, the internal dialogic, the collaborative, and the collective perspectives (48–94; see table of comparison, 52–53). The most individualistic view, which she ascribes (misleadingly, I think) to Plato, treats invention as private and asocial, "engaged in by an individual who possesses innate knowledge to be recollected or expressed, or innate cognitive structures to be projected onto the world" (50). The internal dialogic view locates invention in the thinker's mental conversation with internalized others or social constructs. The collaborative perspective emphasizes how "people interact to invent and to create a resonating environment for inventors" (50), either as partners who invent together or through social interactions that help one person to invent. The collective perspective focuses on the way that individual invention is "encouraged or constrained by social collectives whose views are transmitted through such things as institutions, societal prohibitions, and cultural expectations" (50).

The collaborative and collective views of invention both appear promising for my purposes, and, in fact, I will draw on LeFevre's insightful analysis for later parts of this discussion. But a close look finds that her development of these ideas, and specifically of the collective view, falls short of what is needed here. LeFevre carefully acknowledges the potential of a collective interpretation of invention for explaining how social forces might impinge on the powers of individuals or collaborators to invent, specifically through institutions: "A collectivity infuses its institutions with social facts, and then the institutions, as well as smaller social groups and individuals, operationalize the dictates" (84). But she considers it, at the extreme, nondialectic, implying that "the socioculture itself is what thinks through individuals or by means of individuals" (81). This is an appropriate criticism from her perspective as a specialist in writing and rhetorical invention, interested primarily in how "an individual who is at the same time a social being interacts in a distinctive way with society and culture to create something" (1). But here my goal is to imagine how (or even whether) a social entity (an organization or institution) could be understood as systemically inventive and to ask how such a holistic property would affect or relate to the invention of individuals and leaders in the organization.

In part, LeFevre doesn't offer such an account because she can't find one in composition and rhetoric. In fact, at the time of writing, she noted that the field of writing studies had barely recognized the possibility of collective views of invention; and she called for studies and applications of them to rhetoric (93–94). Ironically, shortly after her study was published the field took a sharp turn toward collective views of rhetoric (in such forms as ideology critique, feminism, Marxism, and identity politics). But invention simultaneously lost its centrality for composition studies and disappeared as an explicit topic; attention shifted from the epistemic role of language to its functions as an instrument of social action, power struggles, ideology, and identity formation.[3] Thus LeFevre's provocative suggestions for examining invention from a collective perspective (for example, doing empirical and historical studies of the ecology of invention) were left undeveloped.

LeFevre foreshadows a systemic or relational understanding of invention in her preferred definition of (individuals') invention as a dialectical process in which "the inventing individual(s) and the socioculture are co-existing and mutually defining" (35). Drawing on Silvano Arieti's idea of the "magic synthesis," she suggests that creativity occurs when "the characteristics of certain people mesh with characteristics made available by their socioculture at a given time and place. . . . A culture cannot 'think' . . . without the synthesis made possible by individuals who interact with culture in certain ways, nor can individuals create ideas in a vacuum removed from society and culture" (36). However, her discussion seems to maintain a separation between the creativity of individuals and the culture that does or does not support or respond to them. Mihaly Csikszentmihalyi takes this idea a step further in his empirical study of ninety-one people who have made creative contributions in a variety of fields. It is his definition of creativity that will best serve my purposes of conceptualizing institutional invention.

Csikszentmihalyi begins by asking not what creativity is but where it is found. He acknowledges that one may attribute creativity to individuals as a set of personal qualities, but finds this definition of creativity too subjective to be useful. Creativity is observable and public only insofar as it "leaves a trace in the cultural matrix" (27). He therefore develops a systemic definition of creativity as jointly constituted or co-constructed in a set of relations among three elements: domain, field, and person. In his initial explanation of these terms, domain refers to symbolic rules and procedures that govern some cultural activity such as mathematics, visual arts, or law. Field "includes all the individuals who act as gatekeepers to the domain. It is their job to decide whether a new idea or product should be included in the domain." Creativity is consummated, one might say, when a person uses the symbols of a given domain to conceive a new idea or pattern that is "selected by the appropriate field for inclusion into the relevant domain." From this perspective, "Creativity is any act, idea, or product that changes an existing domain, or that transforms an existing domain into a new one. And the definition of a creative person is: someone whose thoughts or actions change a domain, or establish a new domain . . . [which] cannot be changed without the explicit or implicit consent of a field responsible for it" (28).

Csikszentmihalyi's approach contrasts with LeFevre's in that rather than interacting with separate "collectives" or their ideologies, the individual, for Csikszentmihalyi, is a part of the system of creativity. The difference becomes visible when we realize that he treats the creativity of famous persons and their inventions as variable over time, depending on their cultural acceptance: "If creativity is more than personal insight and is cocreated by domains, fields, and persons, then creativity can be constructed, deconstructed, and reconstructed several times over the course of history" (30). Csikszentmihalyi proceeds in his book to explore the contribution of each element, how they vary, and how the way they enter into the relation may affect the potential for creativity.

These definitions of terms are not as precise as I would like, and in using them I will refine their meaning to fit Csikszentmihalyi's usage in the study as a whole (and my own needs). For example, domain is best used to refer not to symbolic rules themselves but to the sphere of activity governed by them. Rhetoricians can help greatly to specify the idea of symbolically structured activity as characteristic of a domain (as it does in studies of textual dynamics, genre, and writing in the professions.[4]) Csikzentmihalyi stresses that persons can't contribute inventively to a domain without mastering and internalizing a complex of rules for symbolic manipulations. Rhetoricians would agree, but would stretch the notion of such mastery to cover a much broader set of skills and generic knowledge (of audience, situation, customs, knowledge content, characteristic strategies, epistemological assumptions, etc.). Field seems to denote not just any people who might take up an activity, but the legitimate or authorized participants in the activities of a domain, functioning as an "epistemic court."[5] For example, the field of tennis would not include you and me in our leisure time, but those top-ranked players who are authorized as expert enough to introduce innovations in play and to judge, respond to, use, and disseminate those of others. (That example brings up the interesting possibility—relevant to institutions—of overlapping and intersecting domains: e.g., who is the innovator when a new racquet technology is developed? Who acts as the "field"—the well-known player who uses and advertises the new racquet, the manufacturers who decide to market it, or the public who buys it?)

Csikszentmihalyi tends to focus on domains of discursive knowledge making (i.e., academic disciplines and related professions) or esthetic activity (the arts). But his generic use of "domain" broadens his claims to almost any organized human behavior, and his study (a series of interviews) addresses a great variety of spheres for invention. Examples from the lives of his subjects (see biographical sketches 373–91) include philosophy, acting, physics, sculpture, jazz piano, poetry, zoology, astronomy, social activism, electronics, law, the media, business, philanthropy, politics, and educational administration. Among these are a few administrators and executives whose creative domain is an organization or institution. While Csikszentmihalyi recognizes this possibility, he finds it awkward to call an organization a symbolic domain. "In cases where the responsibility is to lead a group of people in novel directions, work is usually dictated not by a symbolic domain but by the requirements of the organization itself. . . . [Here] the medium is the message; what they accomplish within their organizational structure is their creative accomplishment" (92). He does not develop this idea of an organizational equivalent to a symbolic domain very satisfactorily. Part of the problem is that Csikszentmihalyi does not acknowledge in discussing domains that symbols must structure something, call it "content," and domains differ by the content they structure as well as by their symbolic means and materials. For example, music symbolically structures sound at one level in order to represent human feeling at another (Langer).

For my purposes, I will accept his insight that any sphere of knowledge or cultural activity is symbolically structured and think of an educational institution, like a knowledge discipline or esthetic art, as a domain with its own symbolic rules and procedures, situations, strategies, and generic behaviors. But it might be useful to distinguish broadly between domains that exist specifically and somewhat autonomously to develop knowledge about natural or cultural phenomena (i.e., the traditional academic discipline) and those that are themselves spheres of practical activity. I think this is what is Csikszentmihalyi is getting at in the comment quoted above. As an organization, an institution of higher education is a zone of practice whereby its academic and administrative activities are holistically organized and carried out. According to Zemsky and Massy, responsibilities for the educational enterprise as a whole have increasingly passed from faculty members to others: "administrative and professional personnel ended up holding the institution together—advising students; managing programs, centers, and institutes; and, in increasing numbers, providing the technical support for the faculty members' expanding research efforts" (43). This shift is reflected in strong concerns about the diminished faculty role in shared governance (e.g., Lazerson, Leatherman).

This distinction, if sharply maintained, would get us into a lot of trouble if we tried to account for many of the fields in universities that blend practice and theoretical knowledge (including rhetoric itself), but it may help for now to think about educational institutions as organizations, or spheres of practical activity, that, however, embed domains whose primary activity is to develop knowledge or undertake creative activity for its own sake. It is the tension between the two that complicates any understanding of institutions as inventive.

As an organization, a college or university might be inventive with respect to its own goals and constituent activities or the structures that organize it both literally and symbolically. The two are significantly linked. Barr and Tagg, contrasting a traditional "Instructional Paradigm" with a novel "Learning Paradigm" for undergraduate education, cite management expert Peter Senge in arguing the importance of reinventing structures in educational institutions. They define structures as "those features of an organization that are stable over time and that form the framework within which activities and processes occur and through which the purposes of the organization are achieved. Structure includes the organization chart, role and reward systems, technologies and methods, facilities and equipment, decision-making customs, communication channels, feedback loops, financial arrangements, and funding streams" (18). They see "restructuring" these organizational features as a key to broad educational reform. As I will discuss later, in today's institutions such innovation with respect to the institution itself (its goals, activities, and structures) coexists and comes into conflict with its role and function of housing knowledge domains which are their own, relatively autonomous spheres of creativity.

Csikszentmihalyi's systemic conception of creativity suggests that we might understand institutional invention better by studying the features of complex systems. Theorists of complex systems believe they share common features and qualities that enable unpredictability and novelty. Consider, for instance, Stuart Kauffman's effort to abstract properties of complexity, using the origins of life and the behavior of coevolving species in ecosystems, among others, as case studies, along with computer simulations of abstract complex networks. Kauffman repeatedly cautions that these descriptions at present have only metaphorical application to human systems and culture, while at the same time hoping and believing that there do exist laws of order that govern all self-organizing systems and have everything to do with their ability to evolve to novel, unpredictable states, in what might be called invention. For example, he says,

> The question of what kinds of complex systems can be assembled by an evolutionary search process not only is important for understanding biology, but may be of practical importance in understanding technological and cultural evolution as well. The sensitivity of our most complex artifacts to catastrophic failure from tiny causes . . . suggest that we are now butting our heads against a problem that life has nuzzled for enormously long periods: how to produce complex systems that do not teeter on the brink of collapse. Perhaps general principles governing search in vast spaces of possibilities cover all these diverse evolutionary processes, and will help us design—or even evolve—more robust systems. (157)

I will feel free, therefore, to mine his analysis for suggestive parallels or intuitions about creativity in institutional systems without taking them too seriously as literal claims.

Concerning the origin of life, Kauffman argues that innovation emerges inherently from complexity in self-organizing systems—even a simple chemical system—when they become diverse enough networks of connected parts. In a hypothetical network of chemical reactions, he shows that as the diversity of molecules in the system increases and they become more interconnected, more of the reactions are catalyzed by members of the system until they reach a critical level of diversity (catalytic closure) at which "a collectively autocatalytic system snaps into existence. . . . Life emerges as a phase transition" (62). He explains, "Catalytic closure ensures that the whole exists by means of the parts, and they are present both because of and in order to sustain the whole. Autocatalytic sets exhibit the emergent property of holism" (69).

Imagine a teaching program in which teachers work separately in their individual classrooms in diverse ways to improve writing. If these activities are systemically interconnected through communication and exchange so that they catalyze one another, his theory suggests, at some point the system will pass through a phase transition and become a teaching community, a whole that is capable of institutional invention.[6] One of the problems that faces leaders of such a program is how to balance individual teachers' freedom to experiment and diversify the curriculum (at the extreme, falling into chaos) against the need for coherence and consistency (at the extreme, freezing practice into a rigid order). Kauffman's central insight (illustrated by simulating huge networks of interconnected light bulbs) is that there is an optimal point for innovation in such a system: near the edge of chaos, in the ordered regime: "just near this phase transition, just at the edge of chaos, the most complex behaviors can occur—orderly enough to ensure stability, yet full of flexibility and surprise" (87). Orderly dynamics emerge when the connectivity of the light bulbs is "tuned" so that, like Goldilocks and the three bears, the network is not so interconnected as to produce chaos and not so sparsely connected as to be rigidly ordered (80–81). He speculates that complex adaptive systems evolve to this position: "Perhaps such a location on the axis, ordered and stable, but still flexible, will emerge as a kind of universal feature of complex adaptive systems in biology and beyond" (91).

Another expression of this principle is the idea that complex systems can't find the best solutions to problems but must optimize and compromise. After simulating biological evolution metaphorically as a process of climbing a rugged landscape to achieve the highest peaks (of fitness), he argues that

> both biological evolution and technological evolution are processes attempting to optimize systems riddled with conflicting constraints. Organisms, artifacts, and organizations evolve on correlated but rugged landscapes [i.e., they are both highly interconnected and highly constrained]. Suppose we are designing a supersonic transport and have to place the fuel tanks somewhere, but also have to design strong but flexible wings to carry the load, install controls to the flight surfaces, place seating, hydraulics, and so forth. Optimal solutions to one part of the overall design problem conflict with optimal solutions to other parts of the overall design. Then we must find compromise solutions to the joint problem that meet the conflicting constraints of the different subproblems. (179)

In complex biological and human systems, this picture is complicated by the fact that subsystems (e.g., species in biology or, in an institution, different practical or disciplinary domains) coevolve and thus change the context in which the others evolve. Kauffman argues that such ecosystems over time maximize overall fitness and minimize the rate of extinction: another expression of the idea that systems "may self-tune to a transition between order and chaos" (234–35).

A final insight from Kauffman is the idea that systemic complexity globally escapes mindful human control and planning. Even though we, unlike evolution, can be self-aware of the systems we inhabit and constitute, he suggests that thinking doesn't help that much: we are not much better than Darwin's blind watchmaker, unable to predict the long-term consequences of our actions (243). At the edge of chaos, small moves can unpredictably trigger cataclysmic changes that cascade across the delicately poised system. "All we players can do is be locally wise, not globally wise" (29).

This claim seems to contradict, but is not entirely incompatible with, the ideas of management theorists who want to elucidate the behavior of systems and promote self-conscious systems thinking in order to facilitate the possibility of an innovative culture in organizations. I will use Peter Senge's *The Fifth Discipline* to stand for this large body of work, which documents and teaches principles and practices for organizational creativity in response to complex, rapidly changing environments.[7] Senge's purpose is to build what he calls "learning organizations" by teaching members to practice five "disciplines" that enhance individual and joint creativity. The learning organization is one capable of institutional invention "where people continually expand their capacity to create the results they truly desire, where new and expansive patterns of thinking are nurtured, where collective aspiration is set free, and where people are continually learning how to learn together" (3).

Senge describes complex systems in terms strikingly similar to Kauffman's. For example, he too emphasizes the impossibility of fully understanding and controlling the whole system because of what he calls "detail complexity," the number of variables which "renders all rational explanations inherently incomplete" (365). But his disciplines for systems thinking are designed to help understand and manage "dynamic complexity," the fact that "'cause and effect' are not close in time and space and obvious interventions do not produce expected outcomes" (364). His thesis is that by mastering these disciplines, members of an organization individually and collaboratively can exert leverage, which is to say that they can discern how and where to make small, strategically chosen changes in structure to achieve large-scale, significant changes—echoing Kauffman's comments on tiny changes, but with a more optimistic twist. Similarly, he observes many of the same patterns of innovation as do Kauffman and others, for example, bursts of innovation and rapid learning followed by exponential slowing and limits on growth; but he has practical suggestions for how to recognize and manage them, at least better than the blind watchmaker.

I find it fruitful to juxtapose an understanding of creativity as systemic with a concept of sufficiently complex systems as inherently creative. Together they provide a new metaphorical frame that help us define problems and generate specific questions about institutional invention.

Rethinking Leadership for Institutional Invention

One reason I have chosen these sources to think about changing institutions is their refreshing focus on the inventiveness of a human system rather than exclusively on its function of distributing and controlling power. That shift requires, at least, expanding traditional roles and functions for leaders and, perhaps, radically rethinking the concept of leadership and a good deal else that academics, in particular, take for granted about power, authority, and their relationships to institutions. My goals are more modest here. I simply want to use the perspectives I have introduced to ask more perspicuously how it is possible to enable inventiveness in an institution—and see where that takes me in contemplating new responsibilities for academic leaders.

Let me begin with the fertile concept of "affordance" as defined by ecological psychologists (Shaw and Bransford). Roughly, an affordance is anything in the environment viewed or defined uniquely from the perspective of a given animal (for my purposes, let's say a human being): "the affordances of the environment are what it offers animals, what it provides and furnishes, for good or ill" (Gibson 67). In Gibson's example, the objects we call seats, stools, benches, and chairs afford sitting-on. Other animals of an environment afford "a rich and complex set of interactions, sexual, predatory, nurturing, fighting, play, cooperating, and communicating. What other persons afford, for man, comprise the whole realm of social significance" (68). To simplify here, I will speak of an institutional environment "affording" human inventiveness when I mean what Gibson would call "positive affordance." So I might frame a preliminary question this way: what kind of environment (positively) affords invention?

The theme that some environments afford human creativity better than others is echoed in many of my sources. Csikszentmihalyi exhaustively explores what his subjects find stimulating and inspiring in physical and social surroundings and the ways in which they not only select but reshape environments to afford their own creativity (127–47). More importantly, he regards the domain and the field as offering essential affordances for realizing the creativity of any individual or group (and details how each can provide negative ones). LeFevre uses Harold Lasswell's concept of resonance to explain how social interactions intensify and prolong an individual's inventional activity. Resonance "may occur when someone acts as a facilitator to assist or extend what is regarded as primarily another's invention, or when people are mutual collaborators at work on a task. Resonance also occurs indirectly when people provide a supportive social environment that nurtures thought and enables ideas to be received, thus completing the inventive act" (65). Neither really considers (as many writers on organizational creativity do) how groups or organizations become inventive as a whole (see, e.g., Bennis and Biederman, Douglas); or how leaders might deliberately try to construct organizations or units as resonating environments; or how such environments might work to promote invention.

This was my own explicit goal as a writing program administrator, and I later found Kauffman's analysis an illuminating metaphor for thinking systemically about the problems of creating an inventive climate in a teaching community. Roughly, his theory suggests that to be innovative (flexible and adaptive) a system must achieve an ordered state poised as close to chaos as possible (cf. Tom Peters's concept of "thriving on chaos"). Such a system is highly diverse and optimally interconnected. In human terms, an organization in this state would value risk taking, encourage open communication, and tolerate ambiguity, uncertainty, frequent failure, and mess. Translating these features into an institutional framework means maximizing personal liberty and decentralizing authority and responsibility (for example, to teachers in designing courses, or to students in learning through inquiry or writing on their own topics). Senge describes this approach as one of "local control": "Learning organizations will, increasingly, be 'localized' organizations, extending the maximum degree of authority and power as far from the 'top' or corporate center as possible. . . . Localness means unleashing people's commitment by giving them the freedom to act, to try out their own ideas and be responsible for producing results" (287–88). As Ray Stata (a CEO) explains, traditional organizations have made managers the thinkers and local workers the doers, producing extremes of overconceptualization and pure pragmatism: "The core challenge faced by the aspiring learning organization is to develop tools and processes for conceptualizing the big picture and testing ideas in practice. All in the organization must master the cycle of thinking, doing, evaluating, and reflecting" (qtd. in Senge 351).

However, "balancing" also means remaining within the ordered regime and not slipping over the edge into anarchy—at least if invention is to be comprehensively institutional. In that sense, for invention to be institutional would mean, first, that all its members could participate to jointly change or reinvent it (its purposes and structures) as an organization, and, second, that the creativity of its members would collectively serve not only their personal intellectual goals but also its common purposes as an organization. An alternate concept might be an inventionally chaotic institution with no precise global mission to which its members must subscribe, existing simply to facilitate the autonomous creative work of individuals (or subgroups). Such an institution could in various ways create an inventional climate (bring together mutually stimulating people for conversation—or allow them to coexist in complete privacy and quiet; provide a beautiful physical setting, or an ascetic one; provide financial support or uninterrupted time; and so on). "Institutions" like this exist: for example, research centers, institutes, and think tanks; retreats for artists and writers; and perhaps the kinds of large, complex educational institutions that Birnbaum calls "anarchic" because they house disparate research and teaching communities and are loosely coupled, without superordinate goals or common rules (151–74). As we will see, the tension over innovation in academic institutions lies precisely along the line of this difference. For the moment, however, let me pursue the implications of the more orderly concept of institutional invention.

Senge discusses usefully the problems that localism presents for order (287–301), which are clearly represented in my example of a writing program that has a mission represented in courses collectively taught by a large number of people. In this situation the demands for consistency and fulfillment of common goals conflict inherently with the requirements for inventiveness of individual liberty and responsibility. Senge describes what can go wrong in terms of the "Tragedy of the Commons" and the corresponding need to coordinate creative activities. The Tragedy of the Commons describes the situation, exacerbated by short-term thinking, where "apparently logical local decision making can become completely illogical for the larger system" (294). Within a complex organization that is optimally local and affords creativity by distributing responsibility and control, there is still the problem of coordinating and channeling invention by individuals and local units in the service of the commons and, ultimately, of a collective purpose. These tasks of articulating such a purpose (a rhetorical function) and coordinating the activities of the system define new roles for leaders.

In general, it appears that leadership is a meta-function in systems where individuals and units are inventive and have local control with respect to both the content and means of accomplishing the organization's practical activity. From one perspective, leaders need to take responsibility for the system's capacity to afford invention. It would be impossible to describe such affordances exhaustively (many may be context specific). I have mentioned localism and resonance as affording institutional creativity. My own experience, supported by Kauffman abstractly and Senge practically, suggests that another general affordance is the rhetoricity of the system, the degree to which information is broadly shared and communication is intensive and extensive among members of the system. Bazerman puts this point well in his explanation of textual dynamics in the workplace: "These texts are the transactions that make institutional collaborations possible; they are the means by which individuals collectively construct the contexts out of which intellectual and material products emerge. In the pragmatic worlds of these specialized work communities, texts are a force that transforms human physical and conceptual limits. . . . [T]extual dynamics are a central agency in the social construction of objects, concepts, and institutions" (4). However, Kauffman's analysis of complex systems implies, and my experience confirms, that a system like a teaching community can become too interconnected rhetorically. If everything affects everything else, or if everything must be communicated to everyone before action can be taken, either anarchy or paralysis may result.

Another meta-function of leadership in the inventive organization is, broadly, reflection, which facilitates the ability of the organization to be self-aware and thus to redesign itself. Senge, for instance, describes the essence of the new leader's role as research and design: "What does she or he research? Understanding the organization as a system and understanding the internal and external forces driving change. What does she or he design? The learning processes whereby managers throughout the organization come to understand these trends and forces" (299). Further, leaders as teachers (and coordinators of activity) focus the attention of people in the system primarily on its purpose or vision and the structure of the system: "how different parts of the organization interact, how different situations parallel one another because of common underlying structures, how local actions have longer-range and broader impacts than local actors often realize, and why certain operating policies are needed for the system as a whole" (353). There are abundant resources for understanding reflection in educational contexts and in practice, generally, that might now be brought to bear on considering reflection as potentially an institutional feature, like invention itself, and closely linked to it.[8]

One implication of this discussion is that the traditional idea of a central, authoritarian leader does not survive the change to an inventional paradigm for institutions. If we imagine some of the functions I have described (affording an inventional climate, interconnecting an organization rhetorically, designing and reflecting on organizational structures and processes, and articulating purpose) as new leadership roles, we can also conclude that they aren't very obviously confined to a centralized figure. In fact, one of the principles of a learning organization is that cognition, knowledge, responsibility, rhetorical power, decision making, and systems thinking are distributed broadly—ideally, to everyone.[9] I want to remain agnostic on the question of what specific roles still need to be performed centrally and what special functions still require a single leader for an inventional organization or for its units. I believe that such needs remain and that we must eventually define them to fit the very complicated and conflicted situation of academic institutions, where attitudes toward creativity are deeply entangled with the difficulty of taking or accepting leadership. I turn to that situation next.

Conflict of Visions: Academic Culture and Institutional Invention

It has often been noted that academic culture resists organizational leadership and, more generally, institutional change. Underlying that truth is a conflict between two visions of creativity in relation to academic institutions. William Brown's acerbic analysis of academic values can help us to understand why institutional invention is oxymoronic in traditional academe.

Brown argues that academic culture is governed by a set of normative values, rather than a universal mission or one specific to a given institution. He finds an expression of such values in a summary of "Cartesian principles" (I omit his quotations from Descartes):

1. Intellectual activity should be pursued individually and independently.
2. There can be no limit on inquiry.
3. Rational consideration is universal.
4. The scholar should have objectivity. . . .
5. Personal calculations of a scholar must always be made to amplify the time allowed for scholarship.
6. A secure and stable environment is essential to the cultivation of reason. (18–19)

This ideal of the academic citizen and his or her work (for which he draws on Michael Oakeshot's description of a rationalist) determines that the autonomy of the faculty member, which allows the free pursuit of reasoned inquiry, is a supreme value. Citing J. Baldridge and his associates, Brown describes the university as "a collection of individuals who somehow produce a joint product while operating in a highly independent fashion. All activity is organized around the practitioner's academic discipline. Because of the style of academic performance, the academician is inclined to see all collectives as threats to his free pursuit," including the institution itself (12). Autonomy is thus closely linked to creativity, and creativity is identified with the individual academic, while the institution is organized to make such invention possible: "the essence of the proper academic life is creativity and . . . the profession is organized to allow the participant the time for inventive pursuits" (80). Within the institution, other faculty activities like teaching are assimilated to this model of creativity, while contributions to the organization of the institution ("service") by faculty or others are stigmatized as managerial and noncreative.

These values (which have been described similarly by many other analysts) make it difficult for academics either to accept or to exercise locally institutional authority, to invest themselves in defining or pursuing collective goals of the institution, or even to recognize its organizational structures. They do not, like most employees, view themselves as means to accomplish an end that is the mission of the organization as a whole. Brown argues that "[t]he academician, in effect, hopes to conduct his affairs in an institution that does not really have the attributes of an institution. The university carries the ideal of low power orientation, if not the obliteration of power altogether. . . . Colleagues who seek power . . . are deprecated as local politicians. Administrators who overtly exercise power are viewed with suspicion, and their efforts at rationalizing institutional relationships are opposed" (22). He believes that most academics misunderstand how institutions work as systems of practical activity and are uninterested in the pressures and constraints such organizations work under. For example, they have little idea of scarcity and tend to view resources as unlimited if the goal is deserving (28–29, 32–33).[10] Such a culture is naturally hostile to strong organizational leadership at any level.

This is a view of higher education that makes faculty creativity competitive with institutional innovation (which would displace it and require faculty to direct creative energies toward goals that are neither personal nor autonomous) and even with student learning as itself creative and inventive. Thus, we can argue that if this ideal of the faculty entrepreneur is realized (as it clearly has been at many of the most prestigious institutions), it certainly illustrates one way for institutions to create a climate remarkably hospitable to creativity, although its benefits are limited to a subset of those working and learning there. But if we ask whether this same system of values can support reinvention of institutional goals and structures, or creative action in service of common local goals, or more widespread opportunities for invention (extended to administrators, students, and staff), the answer is no.

This antithesis, however, is not so stark as it appears, and it doesn't refute the systems concept of invention. First, it is not a conflict between individualism and communalism (as Brown overclaims)—in practice, perhaps, but not in essence. Faculty in the most successfully entrepreneurial universities frequently work in teams and groups, labs and research centers, and foster interconnectivity within and across disciplinary lines. If they don't, they still draw on human interactions that catalyze ideas. The current lack of these is, in fact, a source of faculty anomie and represents a breakdown of traditional faculty culture (Massy, Wilger, and Colbeck). Second, faculty do belong to domains and fields (their disciplines) and function within these systems. As I pointed out earlier, educational institutions are domains of practical activity that embed and nurture more purely knowledge-making domains.

The conflict of visions, then, is not a simplistic one featuring an individualistic perspective on creativity that defends faculty autonomy vs. the oppressive collectivism of a bureaucratic regime that discourages invention. It is rather a genuinely difficult conflict between two domains that place rival claims on individuals' inventional abilities and all that it takes for them to be creative in a sphere of activity: time and attention, continual learning, constant practice, systems thinking about the domain, networking, institutional resources, and so on. At issue also is who is recognized or allowed to be creative: in the traditional model, it is only the (tenure-track) faculty, not staff, students, adjuncts and part-time instructors, or administrators. The current environment for higher education exerts pressure on the status quo in both cases: to rebalance the competition for faculty members' creative efforts in favor of local institutional needs, against the exclusive claim of the disciplinary domain; and to broaden support for inventional activity to include students as learners along with those (administrators, staff, faculty in service roles) whose domain of invention is the institution itself.

For a considerable period of our history, certain institutions have successfully enacted (and articulated for others) an idea of institutions acting as "holding companies" (Burton Clark, qtd. in Keller 36) for faculty functioning within autonomous domains and their fields (disciplines). This model, which depended on extraordinary stability and conservatism of institutional structure, was possible during an expansive time when societal goals made funds and popular support available. Now, the balance has broken down because institutions can no longer operate without their own organizational creativity. They need to undertake new activities, reinvent old ones, and transform their structures and processes to adapt to new conditions, make difficult choices, exploit new opportunities, and meet evolving societal needs for education and problem solving. In response, faculty need to grasp and act on their local institution's urgent need for their participation in becoming institutionally inventive. In doing so they must rethink creativity as a more democratic and distributed value. By acknowledging the potential for an institution to become systemically inventive (to be a domain in which individuals and groups can act creatively), faculty could give more weight to collective activities and actions (including teaching) oriented locally and serving an institutional mission rather than disproportionately valuing contributions to their transinstitutional disciplinary domains. Capabilities and opportunities for innovation and creative learning must expand in principle to include students, administrators, and staff in their different spheres of activity, along with the institution itself as a collective. Faculty must also reconsider their negative and passive attitude about academic leadership in the light of new functions and the need for them to take more active responsibilities for the health and collective purposes of the organization.[11]

At the time of Brown's work in the early eighties, he and the other theorists he cited seemed resigned to the view that higher education institutions were essentially ungovernable and needed to settle for some gently modified forms of anarchy. Brown asserted the need to work within both views of creativity, but didn't seem to have very concrete ideas how to do so other than the status quo he described so brilliantly (if not always displaying such keen perceptions for actual practices as he did for underlying norms and ideals). The situation has since evolved to the point that educational analysts are able to describe a new, more collectively oriented academic model which they claim is now replacing the one that Brown, Keller, and others described.[12] One of the most successfully decentralized and entrepreneurial cultures I have observed, that of Johns Hopkins University, recently ran into the limits of this organizational model in undertaking strategic planning for the future. Their report (Committee of the 21st Century) concluded that certain functions like technological innovation needed to be more centrally planned and managed (i.e., in Kauffman's terms, more ordered) while they reaffirmed their commitment to localism and faculty creativity through entrepreneurship (anarchy). In other words, they tried to have it both ways: to rebalance order and chaos so as to remain as close to the edge of chaos as possible, but not quite so close to the point that Birnbaum calls "organized anarchy" (153).

However, faculty resistance remains extremely strong, especially in those institutions that most fully embody the old model. As an issue of fundamental values, the problem can't be solved simply by brute force or through the pressures of external demand. Approaches to this crisis in faculty culture need to address these conceptual incompatibilities, along with the commitments, associations, and responses they engender, and find both reasons and ways to mediate them. The notion of institutional invention developed here may help to imagine reconciliation among the different domains of creativity experienced by faculty members, administrators, staff, and student learners.

My best hope in undertaking this exploration of institutional invention was to start a conversation about the idea that institutions could be inventive, like organisms or academic disciplines, and that we could help to make them so. It's an idea that is novel and alien to many academic minds, though commonplace in other sectors of society. But only those in composition and rhetoric are likely to interpret institutional invention in the context of rhetorical invention and, by extension, of invention and heuristic inquiry generally. We would understand differently, and more richly, the possibility of developing institutional invention as a practical art—and realize that the analogs and interrelationships between invention and rhetoric will make that easier. My references to actual practice of institutional invention have been sketchy, but I hope sufficient to suggest that these abstract theories and models have very concrete and vivid practical translations for someone whose goal is expressly to afford invention within a community of teaching, research, or any other socially organized activity. How can we learn more?

At the exploratory stage of inquiry into institutional invention, one of my purposes was to generate questions fruitful and specific enough to invite further investigation. Attached is a list of questions that grew from this inquiry, ones that I hope others will amplify and pursue.

Second, I suggest that we follow LeFevre's prescription to study examples of institutional invention both empirically and historically, as well as through textual accounts. Academics should look very widely for such examples. For example, in higher education we can look at experiments with teaching and learning communities and project-based courses that use teams to solve problems; study innovative projects on any campus; and learn about experiments in widespread institutional change (such as those on assessment or faculty roles and rewards, among a number sponsored by the American Association for Higher Education). But we can also suspend preconceptions and prejudices to search other sectors of society—business, government, the military, civic groups, religious organizations, charities, nontraditional communities, and others—for ideas, structures, and experiments with organizational invention and related leadership practices that the academy can borrow and adapt to its unique needs and values. Closer to home, I suggest we look very carefully at writing programs as potentially rich and flexible sites, just right in size and human scale for experimenting with innovative organization, coping with (and researching) the human difficulties of change, and trying out new leadership models infused with understandings of rhetoric.

Finally, I would turn to rhetorical theory to further examine and enrich the role of rhetoric within institutional invention (as a means of effecting change; a language for explaining invention; a feature of its operations as a system; and so on). It seems to me, too, that all practical arts inform one another by analogy; so it may be that a practical art of institutional invention will become, in turn, a provocative model for arts of text, rhetorical practice, and teaching.

Appendix

Questions on Institutional Invention

- What conditions enable or define a "climate of invention" for those in a program or unit? How stable can such a state be? Does it run down? Can it be reenergized?
- How do changes in educational institutions play out general patterns of innovation like cycles or learning curves? Do institutional innovations shift from initial, deliberate inventions to self-organized evolution?
- What are the characteristics of an optimally inventive institution or unit? How can we define and sustain the precarious balance between freedom and control, chaos and order—in curriculum, pedagogy, governance, faculty roles?

- How do patterns, principles, and problems of institutional invention apply to classrooms—to student invention and student learning?
- What are some differences between a domain of knowledge making and a domain of practical activity? What can we learn about institutional invention from studying domains that blend theory with practice (artistry)?
- What role do inside forces, conditions and participants play in institutional invention vis-à-vis outside forces, ideas, resources, people?
- Research institutions or labs have been institutionally inventive as collective enterprises: what can we learn from them? How can a teaching group or community be similarly inventive?
- How can we incorporate time into our understanding of the processes and difficulties of institutional invention?
- What is the role of rhetoric in promoting interconnectivity within an organization or unit as a complex system?
- How can we use human-scaled units with a collective mission, like writing programs, to observe and experiment with institutional invention?
- What are the downsides or negative effects of any institutional change? What are the human costs of engaging in and trying to sustain habits and environments of perpetual risk, constant adaptation to change, and continual inventiveness?
- What are the limits of invention as an institutional value? What are some important countervalues?
- What defines leadership in the context of institutional invention? Is there an inherent conflict between creative leadership and widely distributed inventiveness?
- What is the proper role of leaders in changing the institution itself? in fostering a climate for individual invention? Specifically, what are the relations between the creativity or vision of an academic/administrative leader and the creativity of others (faculty . . . staff . . . students) within the institution? For example, (how) might a leader's own invention displace, supplement, enable, direct, generate, or otherwise affect the inventiveness of others?
- How may the goals and forms of reinventing institutions themselves come into conflict with faculty members' goals for creativity in disciplinary domains? What are the tradeoffs between collective invention and individual entrepreneurship . . . between local institutional goals and values and cosmopolitan discipline-based ones?
- What are the obstacles to institutional invention? How, for example, does one invent or mobilize broad-based invention without a sanctioned leadership position or office? from a relatively powerless rank or role? from a position deep within an institution, in a minidomain like a program, department, or committee? against the weight of institutional inertia? in the face of prejudice against one's gender, disciplinary allegiance, color, culture, age, or other difference? How is one's imagination or action constrained by limited resources, peer pressure, the institutional reward system, lack of information, bureaucratic rules, or other factors?

Notes

1. For a sense of this conversation, see the Pew Policy Perspectives that report on the Pew Higher Education Roundtables and related initiatives; and the magazine *Change*, published by the American Association of Higher Education.

2. Compare, for example, definitions of invention by Crowley; Corbett; Young; Young and Liu; and Bushman.

3. Janice Lauer claims that invention was not lost but underwent a diaspora. It has "migrated, entered, settled, and shaped" other areas of rhetoric and composition, becoming "implicit, fragmented" in such sites as writing in the disciplines, cultural studies, and theories of technology ("Rhetorical Invention" 1–2).

4. Examples include: for textual dynamics, Bazerman and Paradis; for genre studies in the academy and other organizations, Bazerman, Freedman and Medway; Berkenkotter and Huckin, Prior, and Swales; for writing in the professions, Duin and Hansen and MacDonald.

5. See (in "Composition Studies") Lauer's adaptation for rhetoric of Stephen Toulmin's term.

6. David Franke is currently completing a dissertation that analyzes how a teaching community functions as such a self-organizing inventional system in the Writing Program at Syracuse University. He uses genre theory to show how that is accomplished through teachers' functional writings (e.g., syllabi, assignments) and interdependent reflections. Although his is not a historical study, I estimate it took about three to five years from the founding of the Writing Program to reach the phase transition where the teaching community became holistically inventive; certainly since its sixth year, it has functioned this way independent of specific leaders.

7. Some other examples of work on creativity, organizations, and leadership include Peters, Green and McDade, Bennis and Biederman, Douglas, Gardner, Kelley, and Bergquist.

8. For work on reflection and reflective practice in educational contexts, see Qualley, Hillocks, Schon, van Manen, Stenhouse, Brookfield, and Phelps.

9. For a practical analysis of leadership at different levels of responsibility in educational institutions, see Green and McDade.

10. For a more current and nuanced study of faculty views on resource limits and their consequences, see Massy and Wilger.

11. Some ideas on expanding the sites of faculty creativity (under the rubric of "intellectual work") are developed in a report for the disciplines of language and literature ("Making Faculty Work Visible").

12. See, for example, Rice on the general shift in faculty culture, and Barr and Tagg for differences between the Instructional and the Learning Paradigms of

undergraduate instruction. Among Rice's points of contrast, he describes the following trends: from "maintaining a primary focus on faculty—who we are and what we know" to "a primary focus on learning"; from "an emphasis on the professional autonomy of faculty" and "highly individualistic ways of working ('my work')" to "increased faculty involvement in academic institution-building" and "greater collaboration ('our work')" (21). When I first heard him outline this paradigm change, I thought it was wishful thinking. Now there are many signs that it is taking place in certain institutions, at least, if not across the board; typically, those under severe pressures—e.g., demographic and financial—or, in some cases, led by a visionary president.

Works Cited

Barr, Robert B., and John Tagg. "From Teaching to Learning: A New Paradigm for Undergraduate Education." *Change* (1995): 13–25.

Bazerman, Charles. *Shaping Written Knowledge: The Genre and Activity of the Experimental Article in Science.* Madison: U of Wisconsin P, 1988.

Bazerman, Charles, and James Paradis, eds. *Textual Dynamics of the Professions: Historical and Contemporary Studies of Writing in Professional Communities.* Madison: U of Wisconsin P, 1991.

Bennis, Warren, and Patricia Ward Biederman. *Organizing Genius: The Secrets of Creative Collaboration.* Reading: Addison-Wesley, 1997.

Bergquist, William. *The Postmodern Organization: Mastering the Art of Irreversible Change.* San Francisco: Jossey-Bass, 1993.

Berkenkotter, Carol, and Thomas N. Huckin. *Genre Knowledge in Disciplinary Communication: Cognition/Culture/Power.* Hillsdale: Erlbaum, 1995.

Birnbaum, Robert. *How Colleges Work: The Cybernetics of Academic Organization and Leadership.* San Francisco: Jossey-Bass, 1988.

Brookfield, Stephen D. *Becoming a Critically Reflective Teacher.* San Francisco: Jossey-Bass, 1995.

Brown, William R. *Academic Politics.* University: U of Alabama P, 1982.

Bushman, Donald. "Invention." *Keywords in Composition Studies.* Ed. Paul Heilker and Peter Vandenberg. Portsmouth: Boynton/Cook-Heinemann, 1996. 132–35.

Committee for the 21st Century, The Johns Hopkins University. Interim Report. Baltimore: Johns Hopkins University, March, 1994.

Corbett, Edward P. J. *Classical Rhetoric for the Modern Student.* 2nd ed. New York: Oxford UP, 1971.

Crowley, Sharon. *The Methodical Memory: Invention in Current-Traditional Rhetoric.* Carbondale: Southern Illinois UP, 1990.

Csikszentmihalyi, Mihaly. *Creativity: Flow and the Psychology of Discovery and Invention.* New York: Harper, 1996.

Douglas, Mary. *How Institutions Think.* Syracuse: Syracuse UP, 1986.

Duin, Ann Hill, and Craig J. Hansen, eds. *Nonacademic Writing: Social Theory and Technology.* Mahwah: Erlbaum, 1996.

Fairhurst, Gail T., and Robert A. Sarr. *The Art of Framing: Managing the Language of Leadership.* San Francisco: Jossey-Bass, 1996.

Franke, David. "The Practice of Genre: Writing as Teaching in an Expert Community." Diss. in progress. Syracuse University, 1998.

Freedman, Aviva, and Peter Medway, eds. *Learning and Teaching Genre.* Portsmouth: Boynton/Cook-Heinemann, 1994.

Gibson, James J. "The Theory of Affordances." *Perceiving, Acting, and Knowing: Toward an Ecological Psychology.* Ed. Robert Shaw and John Bransford. Hillsdale: Erlbaum, 1977. 67–82.

Green, Madeleine F., and Sharon A. McDade. *Investing in Higher Education: A Handbook of Leadership Development.* Phoenix: American Council on Education/Oryx P, 1994.

Hillocks, George, Jr. *Teaching Writing as Reflective Practice.* New York: Teachers College P, 1995.

Kaufer, David S., and Brian S. Butler. *Rhetoric and the Arts of Design.* Mahwah: Erlbaum, 1996.

Kauffman, Stuart. *At Home in the Universe: The Search for the Laws of Self-Organization and Complexity.* New York: Oxford UP, 1995.

Keller, George. *Academic Strategy: The Management Revolution in American Higher Education.* Baltimore: Johns Hopkins UP, 1983.

Kelley, Robert. *The Power of Followership: How to Create Leaders People Want to Follow . . . and Followers Who Lead Themselves.* New York: Doubleday, 1992.

Kevles, Daniel J. *The Physicists: The History of a Scientific Community in Modern America.* Cambridge: Harvard UP, 1987.

Langer, Susanne K. *Feeling and Form: A Theory of Art.* New York: Scribner's, 1953.

Lauer, Janice M. "Composition Studies: Dappled Discipline." *Rhetoric Review* 3 (1984): 20–29.

———. "Rhetorical Invention: The Diaspora." Convention on College Composition and Communication. Phoenix, 1997.

Lazerson, Marvin. "Who Owns Higher Education? The Changing Face of Governance." *Change* 29.2 (1997): 10–15.

Leatherman, Courtney. "'Shared Governance' Under Siege: Is It Time to Revive It or Get Rid of It?" *Chronicle of Higher Education* 30 (Jan. 1998): A8–9.

LeFevre, Karen Burke. *Invention as a Social Act.* Carbondale: Southern Illinois UP, 1987.

MacDonald, Susan Peck. *Professional Academic Writing in the Humanities and Social Sciences.* Carbondale: Southern Illinois UP, 1994.

"Making Faculty Work Visible: Reinterpreting Professional Service, Teaching, and Research in the Fields of Language and Literature." Report of the MLA Commission on Professional Service. New York: MLA, December 1996. Rpt. from *Profession 1996.*

Manen, Max van. *The Tact of Teaching: The Meaning of Pedagogical Thoughtfulness.* Albany: State U of New York P, 1991.

Massy, William F., and Andrea K. Wilger. "Improving Productivity: What Faculty Think about It—and Its Effect on Quality." *Change* 27.4 (1995): 10–20.

Massy, William F., Andrea K. Wilger, and Carol Colbeck. "Overcoming 'Hollowed' Collegiality." *Change* 26.4 (1994): 10–20.

Peters, Tom. *Thriving on Chaos: Handbook for a Management Revolution.* New York: Knopf, 1987.

Phelps, Louise Wetherbee. "(Re)Weaving the Tapestry of Reflection: The Artistry of a Teaching Community." *Rhetoric Review* 17 (1998): 132–56.

Prior, Paul. *Writing/Disciplinarity: A Sociohistoric Account of Literate Activity in the Academy.* Mahwah: Erlbaum, 1998.

Qualley, Donna. *Turns of Thought.* Portsmouth: Boynton/Cook-Heinemann, 1997.

Rice, R. Eugene. "Making a Place for the New American Scholar." Washington, D.C.: AAHE, 1996.

Rudolph, Frederick. *Curriculum: A History of the American Undergraduate Course of Study Since 1636.* San Francisco: Jossey-Bass, 1977.

Senge, Peter M. *The Fifth Discipline: The Art and Practice of the Learning Organization.* New York: Doubleday, 1990.

Shaw, Robert, and John Bransford. *Perceiving, Acting, and Knowing: Toward an Ecological Psychology.* Hillsdale: Erlbaum, 1977.

Stenhouse, Lawrence. *Research as a Basis for Teaching: Readings from the Work of Lawrence Stenhouse.* Ed. Jean Rudduck and David Hopkins. Portsmouth: Heinemann, 1985.

Swales, John M. *Other Floors, Other Voices: A Textography of a Small University Building.* Mahwah: Erlbaum, 1998.

Young, Richard E. "Invention." *Encyclopedia of Rhetoric and Composition: Communication from Ancient Times to the Information Age.* Ed. Theresa Enos. New York: Garland, 1996.

Young, Richard E., Alton L. Becker, and Kenneth L. Pike. *Rhetoric: Discovery and Change.* New York: Harcourt, 1970.

Young, Richard E., and Yameng Liu. Introduction. *Landmark Essays on Rhetorical Invention in Writing.* Ed. Richard E. Young and Yameng Liu. Davis: Hermagoras, 1994. xi–xxiii.

Zemsky, Robert, and William F. Massy. "Expanding Perimeters, Melting Cores, and Sticky Functions: Toward an Understanding of Our Current Predicaments." *Change* 27.6 (1995): 40–49.

Conflict in Community Collaboration

LINDA FLOWER
JULIA DEEMS

Urban community groups are intensely rhetorical forums. They are a fitting place to study what classical rhetoric called the art of invention—the heuristic process by which people arrive at probable knowledge by posing problems, naming conflicts and questions, and building a persuasive case (Enos and Lauer 79–80), and where the goal is to arrive at substantive arguments. Community groups give us a window on the deliberative process of attempting to build a consensus that will allow people to construct productive knowledge for social ends. However, when the topic is the troubled relations between low income landlords (trying to maintain old buildings on limited resources) and tenants (trying to live on uncertain incomes) in inner city communities, these grassroots dialogues are not a place to look for easy consensus. Shaped by poverty, racial tension, a crumbling urban infrastructure, and local social history, the landlord-tenant problem admits no easy answers. It is a prime example of an issue that can not be resolved by a technical art or science. It stands in the province of rhetoric, as Aristotle defined it, as one of the "things about which we commonly deliberate." Although such discussions seem quite distant from the tradition of deliberative rhetoric prized in academic forums, the problems they pose stand as open questions. They call for the reasoned deliberation Aristotle describes, in which rhetoric is not reduced to the mere persuasion of others present, but functions to discover "the available means of persuasion in a given case," to mount the arguments that best justify decision. As Plato predicted, debates often turn on those disputed ideas and terms (on which "the multitude is bound to fluctuate") that would seem to call for systematic analysis, or the dialectic of division and collection Plato urged for getting at the heart of the matter. While Plato argues that the systematic analysis that leads to an understanding of central values is important, however, as is true here, consensus itself can be problematic. Communicators often need to act as if they have reached agreement, but often the consensus is not about ideas but is instead about the agreement to act, to move forward. Often "[communicators] have to assume that they could, in principle, arrive at an understanding about anything and everything" (Habermas 150). Often it is this willingness to assume consensus, rather than actual deliberated consensus that moves the group forward. In practice, this deliberative process is often short circuited. In the face of problems they cannot solve, community groups invoke a discourse of complaint and blame and come to depend on an oppositional rhetoric which invites an advocacy stance from members. Here is how one long-time community developer describes it:

> And I [have] attended a number of those meetings and there was just a group of landlords just trading horror stories. . . . 'Cause one of the big problems with the tenant, or the landlord meetings is they have come in, for two hours they talk and nothing, nothing ends up at the other side. . . . [They leave thinking] I feel better tonight and I go back for a week and then I come back next week and I talk again and still have the same feelings and [I'll] still be in the same place, but I'll feel better 'cause it's off my chest [And essentially you're talking to your own] fears. They've got the same problem, not the people to help you solve your problem. . . . There's no text, there's no decision, and if there is a decision, the decision is that they all agree that they still feel the same way. (Kirk, Final Interview 8–9)

Another community organizer bemoans the evanescent nature of the conversational understandings that do develop: "In our groups . . . we will argue about a topic for discussion, a situation, one month, and they'll come to some kind of a consensus or agreement and then a month later, they'll all forget, nothing will happen, and they'll argue about it again" (Dave, Final Interview 12).

And another suggests why commitment alone is not enough: "I'm a very active participant in my community. . . . Everybody wants to go there and be there one hour and get it, everything accomplished. But the funny thing is that we never get beyond the issue that you wanna talk about because (laughs)—because people . . . people oftentimes come with their own agendas. . . ." (Liz, Final Interview 34).

While the conflict between hard-to-call competing claims (such as equity versus community) calls for a balanced, even dispassionate consideration of alternatives, where debate can take the place of force (Perelman and Olbrechts-Tyteca), this dispassionate consideration is not the norm in community meetings. In this case, then, what can invention offer?

Deliberation amidst Diversity

Although this situation seems to cry out for a more robust rhetoric of deliberation and consensus building, in this context the language of Aristotle, Plato, and Perelman has an air of book learning. And conventions of scientific, technical, or bureaucratic discourse (that could no doubt structure this discussion more efficiently) are not an adequate alternative. One reason is that this is not the discourse of a homogenous group—voting Athenians, a New England town meeting, a legal or academic forum. The dialogue of the inner city must operate in the context of cultural, economic, racial, and educational difference. It must recognize not only competing interests but the alternative discourses people bring to these discussions, from legal assertion, to personal narrative, to the rhetoric of social justice. In authorizing difference, many grassroots groups cultivate a multi-voiced discourse which refuses to privilege the discourses of the technocrat, bureaucrat, or academic. Living on the margins they identify themselves with voicing rather than suppressing conflict and with an adversarial stance toward institutions of power. Unfortunately, this stance also tends to support an oppositional discourse of complaint and blame that is better adapted to voicing conflict than exploring ways to resolve it.

In the face of this history, urban communities face a growing need for an alternative rhetoric that is generative and openly deliberative rather than adversarial. As the budgets of urban centers shrink and the sense of shared civic responsibility for cities decreases, much of the decision making that used to be centralized in city and county agencies is being transferred to neighborhoods. Although these changes give people more control over their own lives, along with that power comes the burden of action and the cost of failure. At the same time, this alternative rhetoric must speak the language of grassroots groups; it must be a rhetoric that can articulate difference, put conflict squarely on the table, and let multiple voices, which do not share a common discourse, have a place in defining and resolving problems. Inner cities need a discourse of both deliberation and diversity.

This paper is in part an account of a community experiment trying to address the contested issue of landlords and tenants through explicitly rhetorical strategies for planning and deliberation, organized around the (quite unusual in this setting) practice of collaborative writing. Initiated by an urban settlement house and its Community Literacy Center (CLC), this five-session collaboration between a small group of landlords and tenants was designed to begin in conversation and end in a useful text. For the CLC this project was also a maiden voyage into housing issues, designed to explore how its literacy-based alternatives to the discourse of advocacy and opposition would fare in such a forum. Like the Center's other projects, discussion was structured by the practice of collaborative planning, which meant that each member of the group was committed, on the one hand, to articulating conflict—vigorously representing a competing perspective on inner city landlords or tenants—and on the other, to supporting and developing each other's position in planning and writing a useful document.

For us as researchers, the focus of inquiry was on conflict and on how this community collaboration, designed to bring troubled issues up for deliberation, handled difference. The CLC project offered a chance to track a process dedicated to the intentional articulation of conflict and to ask: How is such conflict negotiated when writing enters the picture and collaboration is structured around rhetorical planning? After a brief look at the social context, methods, and people involved, we will argue that the inventional process we observed was not a consensus-building process, but a constructive one which gave rise to the active strategies for negotiating the conflicts the process raised.

The Context for Conflict: Bringing More Voices to the Table

The Community Literacy Center is a collaboration between Community House, an eighty-year-old landmark of Pittsburgh's Northside, and the Center for the Study of Writing and Literacy at Carnegie Mellon. It is helping to reinvent an older tradition of community/university collaborations begun in turn-of-the-century settlement houses like this one and Hull House of Chicago, where the problems of urban neighborhoods drew university faculty into a combination of inquiry and grassroots activism. As a grassroots lab for social change, the CLC argues that change can come through education, collaboration, and writing that lets people make their voices heard. It builds its educational vision on the theoretical base of cognitive rhetoric, focused on the problem-solving strategies people bring to problem analysis, collaboration, and argument. For five years, the CLC's projects had helped inner city teens produce documents and public community conversations on issues such as risk, violence, and school reform, working one-on-one with a mentor from Carnegie Mellon (see Long). The relationship between mentors and teens was structured around the rhetorical practice of collaborative planning, in which the teenage writer holds the role of Planner and the college student takes the role of Supporter. In planning and writing sessions the Supporter helps the writer to develop and articulate his or her own ideas, by offering not only social support and acting as a sounding board but by prompting the Planner to think rhetorically in terms of a key point and purpose, the needs and possible response of readers, and the range of text conventions that might support purposes or work for given readers.[1]

The landlord and tenant project was the beginning of a series of new projects called Argue, working with adult community-planning groups and focused on building document-based plans and arguments for action. Although they were structured around collaborative planning, the one-on-one practice that had been used in school settings needed to be transformed into a group practice that supported not only collaborative planning but a collaborative text. In this project the CLC literacy leader, Lorraine Higgins, became both the Supporter (prompting the group to consider key points, purposes, audience, and text) and the facilitator. That is, after a short training session on strategies for planning and supporting, she recorded developing and conflicting plans on a chalkboard as people talked and reminded members of the group to take over the task of supporting and prompting one another. Higgins's (1992) own research on the construction of argument had explored contrasts between the rhetoric of inquiry valued in the university and the rhetoric of opposition and advocacy valued in urban communities. The goal of Argue was to bridge these discourses—to build community-based plans for concrete action, but at the same time to construct these plans in an atmosphere of inquiry that could lead to new solutions by bringing some typically marginalized voices to that table not just as advocates but as collaborators in a solution.[2]

The CLC obviously played an important role in shaping the collaboration we studied since it is the CLC facilitator who structures discussions as collaborative planning sessions and moves the group toward the production of text. However, the overriding goal of this CLC project was to bring more voices to the table, to structure discussions in which opposing positions were not only solicited but supported. This process also created a context for conflict, opening the door to more direct negotiation. Although the interpersonal conflicts that fire up the oppositional discourse Kirk describes may be discouraged in collaborative planning, substantive conflict around opposing perspectives is encouraged. And once these positions are on the table, the need to write forces the group in some way to deal with them. So although this is not a study of a "typical" community discussion of the sort Kirk describes, it revealed some ways of building an argument in the face of diversity that differ from the patterns of normal academic argument. And it suggests that both writing and educational strategies such as collaborative planning can play a positive role in community settings.

Portraits of Participants

Because we wanted to bring the conflicts from the community to the table, each person asked to participate in this project had had experience as either a landlord or a tenant (frequently as both), had been involved in the community debate on this issue in the past, and often had engaged professionally in some area related to landlord-tenant interactions. Additionally, they represented a range of socioeconomic and educational backgrounds. This group was comprised of:[3]

DAVE. As the full-time, paid president of a local community group and a community organizer, Dave Rice worries about the effect individuals have on the community as a whole. And while he recognizes the value of community groups, he believes that often the leader of the group has too much control over the decisions that are made—frequently because members get "volunteered" to research issues or set meetings and then are not willing to do the work required.

KIRK. Kirk Murphy is a member of a small, grassroots community development organization that replaces vacant lots and boarded buildings with affordable housing in an inner-city neighborhood. Part of his work involves motivating landlords and tenants to maintain buildings, keeping key corridors alive and attractive to small businesses and potential home buyers. One of the problems between landlords and tenants, he argues, is getting individuals to take responsibility for their actions.

LIZ. Liz Marino, a mother of four in her early thirties, is very active in her community council and on school committees and is known in her diverse Pittsburgh community for being an energetic and fair mediator in landlord-tenant disputes. Marino has an unflinching commitment to making her mixed urban neighborhood "work" but admits that community meetings can be discouraging at times, especially when people come in with their own agendas.

LUWANDA. LuWanda Baker is a single, African American mother employed in a pharmacy. LuWanda, a tenant who had moved "ten times in ten years," brings with her a range of experiences, from dealing with an absentee landlord to participating in a subsidized rent-to-purchase program.

Because these four knew at least one other member of the group and had occasionally worked on community groups together in the past, they shared some understanding of the history of their community. At the same time, they also had their own values and beliefs, and these affected their attitudes and actions as they participated in this discussion.

Tracking a Community Collaboration

The Argue project was designed to meet for four sessions during which the four participants would articulate and explore the causes of landlord-tenant conflicts while representing either a landlord or tenant perspective (as opposed to articulating only their personal beliefs). They would use their analyses and discussions to write a Memorandum of Understanding that would not only fairly and accurately reflect the conflicts but would advance community thinking on these issues. In the spirit of community activism, this document was not to be an end in itself, a mere exercise, but a useful tool for action that the group would decide to take. (In the end the group elected to meet an additional time to complete the document, and two of its members helped produce a subsequent booklet and community conversation that involved housing groups around the city.)

In session one the group received an overview of the collaborative planning method, then with prompts from the facilitator opened their discussion of the major issues and conflicts that landlords and tenants face. In session two, the group was prompted to explore conflicts further, to develop a purpose and audience for their memorandum, and to consider text conventions they might use—and to take responsibility for writing small portions of the text between meetings. As this planning continued into session three, the group was encouraged to support individual writers as they overviewed their plans and ideas for their section of the memorandum at the table. In the last two sessions, the group jointly read drafts and gave revision suggestions, with prompting to support each other as writers and to consider different perspectives on the issues.

Each session was recorded both on audiotape and videotape (with a stationary video camera). In between selected sessions, each writer taped a self-interview, responding to questions about the goals/expectations/ ideas they had going into the session and how they saw their goals faring. A final interview, conducted by Flower, moved from open-ended requests for participants' evaluation of the process to direct questions on how they saw conflicts being addressed and negotiated in the sessions.

People brought various motivations to their participation in the sessions. The four community members who came to this project were motivated to find new ways to get something done on an issue they cared deeply about, as we have seen in their profiles. Flower sat in on the sessions not as a researcher, but as President of the CLC's Board, who with the group was asking whether the CLC's writing-based, educational agenda and collaborative strategies could work in this adult, community development context.[4] In the spirit of helping explore this question, the participants agreed to tape self-interviews and reflections between sessions in order to evaluate how this CLC project was working for them and the community. Therefore the research questions we pose here about conflict and collaboration were on the table as public questions shared with the participants. The process of collaborative planning itself asks writers to reflect on their own process and to develop the metaknowledge that leads to strategic choice; therefore, the reflection and evaluation that contributed to our inquiry were a normal part of the process under study.

To build a more in-depth picture of how significant conflicts can be negotiated over time, we tracked the fate of two sustained areas of disagreement over the course of this project. One was an early point of contention over the "disputed term" of process which is central to the group's solution. The second conflict, a central disagreement over how to define the problem of landlords and tenants itself, remains unresolved at the end of the project, even though the group agrees on a final text.

Our analysis of these conflicts is based on transcripts of all five sessions, the self-interviews, and the final interviews, as well as the texts and drafts. This allowed us to conduct a strategic analysis of these conflicts. In order to understand the internal and interpersonal negotiations that let people construct meaning, we argue that it is not enough to analyze moves, actions, or strategies alone without understanding why people are taking that action, without insight into their strategic knowledge: that is, their goals and awareness as well as observable strategies. Our strategic analysis then attempted to identify (or make reasonable inferences about) the reasons or goals behind the moves people made, to document the strategies/moves they used, and to seek evidence about the degrees of awareness and sense of options they brought to this process. A strategic analysis raises, of course, problems of evidence since goals and awareness are typically much harder to document. On the other hand, we would argue that the knowledge that matters most in such collaborations is the more complex strategic knowledge that guides internal and interpersonal negotiation.

The Negotiation of Collaborative Meaning: Consensus and Construction

The CLC creates a forum for collaboration that puts the substantive conflicts surrounding open questions up on the table. It also poses an interesting theoretical question about the goal of such collaboration that we wish to broach at the outset. Should we envision this as a process of building consensus around shared meanings, or of building meaning itself? Some accounts of collaboration—and its virtues—focus on the social process by which people arrive at consensus, the way in which belief takes the status of knowledge by becoming socially justified in a community of peers (Bruffee; Latour and Woolgar). In this picture of social consensus building, conflict is a generative force that introduces new beliefs or ideas, around which a new consensus can form, through the power of argument or perhaps just the power of power. Likewise, the role of rhetoric is "to aim chiefly at reinforcing communal values, strengthening adherence to what is already accepted" (Miller 85). And that, of course, is also the first problem with consensus, which leads critics like John Trimbur to argue the place for dissensus: if your position wins the contest for social acceptability, my more marginal voice, or less conventionally justifiable position, may lose in its bid to become "knowledge." The tradition of rhetoric, from Plato's dialectic interrogation of competing truth claims, to the zero-sum game of high school debate, seems to support this competitive view of knowledge construction and democratic consensus (may the best idea win). However democratic consensus does not guarantee that the best ideas will gain force (Rescher). In fact, consensus may reinforce existing arguments and values rather than identifying and elaborating on new and unheard positions. As a consequence, the common call for consensus may not offer the best solution.

Moreover, the knowledge in question tends to exist as a set of propositions, positions, or beliefs. But what if the goal of collaboration is to bring marginalized voices (without fully articulated positions) to the table or to support the discourse of those who traditionally lose the contest for public justification? Consensus building around the most "justifiable" position may not be the most desirable goal for collaboration. In arguing for dissensus, Trimbur sees collaboration as "not merely a process of consensus-making but more important as a process of identifying difference and locating these differences in relation to each other" (610). The goal is not "an agreement that reconciles differences through an ideal conversation but rather . . . the desire of humans to live and work together with differences" (615).

The importance of consensus (defined as a shared, collective sense of a group's experience) has also come under question in organizational theory. Do groups have to achieve consensus in order to take action? "One theory is that organized action is the product of consensus among organizational participants, a view that has led to the conceptualization of organizations as systems of shared meanings. . . . A second view . . . argues that only minimal shared understanding is required, because organization is based primarily on exchange (e.g. of work for pay)" (Donnellon, Gray, and Bougon 43).

Anne Donnellon, Barbara Gray, and Michel Bougon argue, however, that organized action can occur in the absence of shared meaning when there is a repertoire of communication forms that "allow members to coordinate their actions." Their discourse analysis of a organizational conflict shows how group members who held competing interpretations of an event used discussion to arrive at multiple routes to the same end or action, without ever reaching a consensus or a shared understanding of their joint experience. Collaborative action, this work suggests, does not have to depend on shared belief, identification, or consensual meanings, if people can communicate their way to a common organized action. Rather than relying on consensus building to achieve the work of moving people to action, then, perhaps we need to consider how invention can support dissensus—not as an end in itself, but as a tool to focus the discussion on the end goal rather than the contentious terms. Rather than seeing invention in the divide-and-conquer view of Plato or as the dispassionate systematic analysis of Chaim Perelman and L. Olbrechts-Tyteca, perhaps invention for a pluralistic rhetoric should support the persuasive appeals to move people to action.

In our study, achieving social consensus—defined as a shared representation that could claim the socially justified status of knowledge—played only a limited role in the way this group moved to text or to action. Instead of trying to win social justification among competing positions, instead of trying to build consensus around a selected proposition, this collaboration was a construction process in which people responded to conflict by constructing new meanings and a plan of action. Conflict, we will emphasize, did not evaporate in the light of happy consensus; people came and left with strong competing representations of reality and response. What they constructed was not a shared definition of the problem, but a literate action—a text that represented their willingness to move forward with their project. More importantly, we will argue, the rhetorical process of structured collaboration and its drive to text (like Donnellon's communication strategies), let this group articulate and maintain independent perspectives and still build a representation on which they could act. Here invention served to move the group forward, not by focusing on all arguments but by allowing the group to focus on those arguments about which some agreement could be formed.

Does this distinction between consensus (around a preexisting proposition) and construction (of a new multivocal meaning) matter? At some level of analysis, of course, any form of agreement is a social consensus. What does it matter if knowledge is being made, not just promoted? And what if individuals actually hold strikingly divergent personal representations from one another or from those representations that claim the status of public knowledge, if dissenters fall quiet and fall in line for the vote? We will argue that unless we account for individual meaning making within a collaborative process—for the resistant, unreconstructed, unassimilated representations of individual writers—we are likely to create a reified notion of knowledge that no one really holds. This blind spot to personal representations and nonconsensual knowledges presents an obvious problem if one is teaching individual students. It can be a mistake to assume students "learn" what you intended to "teach." But a purely social view of public knowledge also sets us up to misunderstand the dynamics of collaboration in communities where flashes of apparent consensus turn out to be flashes in the pan, and positions with the apparent public status of "knowledge" regularly fail to elicit supportive action. A generalized account of social construction will tell us which ideas get repeated over time. But a more closely observed account of writers' social cognitive processes can tell us how writers privately and jointly construct meaning. Secondly, a social cognitive account of how writers use collaboration to construct new, negotiated meanings throws light on the way people deal with conflict. Carolyn Miller (1993) has called this the challenge of the new rhetorics: to develop a "rhetoric of pluralism [that] must speak not only to the diversity within any given community but also to the diversity of communities that coexist and overlap each other" (91). Looking at collaboration as a constructive process reveals the (to some, surprising) role writing and rhetorically based strategies for collaboration can play in community deliberation.

Negotiating Conflict in the Construction of Meaning

Meaning making, whether in the mind of an individual writer or in collaboration, is often a constructive response to conflict. The model of collaboration as consensus building which we have been questioning places people in the midst of competing propositions and beliefs vying for adherence. The model of negotiation we propose places writers within the midst of multiple social, cultural, and linguistic forces (Flower, *The Construction*). Such forces range from personal goals to literate conventions, to the expectations of an audience or the pressures of a collaborator. This array of "outer" forces (or more accurately those forces that gain a writer's selective attention) gives rise to a set of inner voices that enter the writer's thoughts and would shape meaning in their own image. Such voices not only offer language and concepts, but they also urge priorities, whisper caution, demand the limelight, or propose structure. And, critical to our case, these voices also come in conflict with one another as they introduce competing attitudes, values, and bodies of knowledge, as well as the alternative strategies for persuasion and multiple social expectations this rhetorical situation calls into play.

In the collaboration we study, some of the forces whose meaning-shaping voices are most visible in negotiation involve:

- the social context of this event, from the neighborhood's long history of interracial relations and activism, to the more immediate social goals of the CLC, to the practice of collaborative planning which structured social interactions, and the (unusual) expectation that each member of this community group would produce text;

- the personal representations of the landlord-tenant problem that each member brought to the table (and was expected to speak for);
- the shifting personal and power relations among the people at the table;
- the various conventional discourses (from legal advice to personal narrative) that introduce alternative sets of conventions and expectations into the discussion. In addition to this heteroglossia of conventions, the immediate discourse created its own set of repeated claims, metaphors, and words that imported other histories to the discussion (regardless at times of the speaker's intended meaning);
- finally, the strategic knowledge of individual members came into play, that is, their personal goals, the strategies they brought to collaboration, and the awareness they had of their own moves and options.

Negotiation in this constructive process is not like a union arbitration, giving up *x* amount of income for *y* amount of security. Negotiated meaning making occurs when writers rise to awareness of competing voices and build meaning in response to that conflict, without trying to identify all of the arguments or even the strongest arguments. Even if that awareness is momentary, it can produce a new understanding that acknowledges competing goals and constraints. At times writers negotiate conflict in the sense of arbitrating among power relations, choosing what voices to hear, what to deny. At other times negotiation is a form of embracing multiple conflicting goods in the sense of navigating a best course, shaping a meaning to honor as many values/voices as possible.

Insofar as a practice like collaborative planning can influence the way writers deal with conflict, the process we want to foster is one in which writers construct a negotiated meaning, rising to greater reflective awareness of the multiple voices and sometimes conflicting forces their meaning needs to entertain. The understandings writers come to in a text are a provisional resolution constructed in the middle of both an internal and a face-to-face conversation. Such negotiation is not "giving in" or settling for less, but reaching for a more complex version of a best solution.

Our account of how this community group negotiated conflict starts with what may be a common but little recognized feature of such collaborations—the creation of an apparent consensus. As this chimera of agreement falls apart (when the individual representations of a conversationally "shared" idea emerge), we will observe two ways groups deal with genuine conflict—by finding consensus in action and by a more subtle, strategic process of transforming the group text.

The Problem of Misleading Consensus

In community groups, members come to a discussion, voice their views, negotiate with others, and walk away feeling that they have come to some agreement or resolution. Yet this consensus may be tenuous at best and frequently will not last. Not surprisingly, when that apparent consensus breaks down, group members may feel frustrated and even betrayed. This common scenario raises two questions: Why is this consensus so often fleeting? And why do people assume it is so necessary to achieve?

The collaboration we studied was no exception. And as we observed repeated occurrences of apparent consensus unraveling, we began to see how the goal of consensus could itself be misleading. Community activists often describe their work as trying to galvanize a community into agreement on an issue. But this attitude, which motivates grassroots political activity, suggests that the public ought to have a shared vision, and that by drawing individuals into a shared vision, the ground swell will lead to action. In spite of this attraction to working on a shared mission and the belief that it is a precursor to success, the desire for sustained consensus and belief that it can exist may set up unrealistic expectations. The expectation that a diverse inner city group ought to achieve consensus on goals, for instance, is often unrealistic, yet when it is not met, the ideal of consensus would force us to conclude that the group has failed. But perhaps it is that expectation itself which is at fault.

In order to understand why consensus broke down in this community group, we began to look at the conflicts underlying moments of apparent consensus, focusing on the strategies, goals, and awareness held by the individuals moving towards consensus. It became clear that individuals in the group were bringing to these moments of conversational consensus radically different interpretations of the common topic. Instead of seeing consensus as a moment of simple agreement, we began to see these points of apparent consensus as sites of negotiation among conflicting representations.

The following analysis, focused on a moment of apparent consensus, attempts to identify the conflicts embedded in the two issues under discussion—the purpose of the document and the process (involving landlords and tenants) it was going to support. Conflicting goals, we saw, did not always surface immediately. Individuals do not explain their own goals to the group, and group members typically do not seem to infer that their goals may not be shared. Group members do appear willing to come to consensus—in fact, they act as if they have reached consensus. The attitude of the group is congenial, the tone is relaxed, and all four of the participants are actively engaged in negotiating the purpose of the document. But instead of hearing only consensus, it is possible to hear both consensus (about the concept of "process") and dissensus (about what form that process ought to take, how it ought to be defined, and its features):

Liz: —let's say the purpose of the document would be to, develop a process by which we can—
LuWanda: Get a better understanding between the landlord and the tenant.
Liz: —of the expectations—
Dave: Of each.
Liz: —let's get that word in. The expectations—
Lorraine: Do what with the expectations?
Dave: Before, during, and after the tenancy.
Liz: Yes.
LuWanda: [inaudible 300]
Dave: Cool.
Liz: And the relationship.
Kirk: The process of clarifying—
Liz: Let's get this in [inaudible over others talking 302]
LuWanda: The ongoing relationship, not just the entrance relationship. Ongoing throughout the term of the lease.
Dave: And after the exit. (1.45)[5]

If we were to analyze this linguistically we might see this as an "exchange," composed of an "initiation and any contributions" where later utterances show "compliance" to the minitopic being addressed (Stubbs 135). As such, it might be seen as "an accumulation of shared meaning" (116) where the accumulation occurs around the idea of process.

Other evidence supports this image of a developing consensus. The consensus can be seen, for example, to extend beyond defining the relationship that ought to exist between landlords and tenants and to also establish the purpose of the document. In this agreement on process and purpose, we see a comfortable informality, a responsiveness, a desire to be part of the dialogue. The group members listen to and affirm one another: Dave listens to the others and responds with "Cool"; Liz listens to Dave and responds, "Yes." Later adding "[L]et's get that word in" and "Let's get this in," Liz stands in for the group implying "we should." Through this collective speech act (160), Liz not only establishes her own goals ("let's say the purpose is") but also draws the group into her own objectives.

The highly cohesive nature of this passage further suggests that a shared representation exists. Repetition ("ongoing"), additiveness ("and"), negation used as a tool to set apart ideas ("not just the entrance relationship"), and shared terminology (for example, the repetition of the words "expectations," "relationship," and "process") (Halliday and Hasan) all support this cohesion: not only are the words friendly, but the tone is as well. The group wants to share meaning: they finish one another's sentences and build on the ideas of others. In doing this, every voice is heard (and to some extent accommodated) within the discussion. The idea of what the process ought to be is still being shaped (listen to it move from "understanding" to "expectations" to "relationships" and "clarification"), but nevertheless the tone of this session is confident; the group members are excited about the level of consensus they have managed to achieve. This must be the kind of moment that community workers look forward to most. In this moment, the group members seem confident that their ideas are being affirmed and that this is a signal of the group's essential agreement.

In spite of these signs of a shared consensus, however, this passage can also be read as evidence that group members hold wildly different goals and representations. If, instead of a conversational analysis, we conduct a strategic analysis that tries to construct purposes and ideas underneath conversation and behind the notion of process, then the signs indicate a lack of agreement. After analyzing the process in which they are engaged, it seems easy to predict that the consensus reached in this moment will inevitably break down, as, in fact, it does. Let's turn to the sessions to hear how these different goals and representations are embedded in the group's discussions.

In retrospect, Liz's and Kirk's representations of the purpose of the document (whether to establish a process or to inform community members about already existing processes) seem to be at odds. While we cannot demonstrate that the group members are aware of their own divergent interests (they rarely, if ever, comment on their disagreements in the sessions), it is clear that Liz, with her interest in mediation, thinks the group ought to develop an explicit process that can be used to teach landlords and tenants what their responsibilities are and hopes a renter's checklist will be "incorporated into the process" (3.14). Kirk, however, believes "[a renter's checklist] actually should be part of a lease" (3.14). Likewise Kirk, in his desire to have landlords and tenants accept a process, seems aware of social and historical constraints working against such a process (and thus wants to establish a rationale for convincing landlords and tenants) (3.23), while Liz seems confident that landlords and tenants will implicitly accept that a process is necessary. By the last group session, Kirk argues that "really the purpose of the document was only to investigate . . . a process or what could be done" (4.25), but Liz has a different view—she argues that the group has agreed to offer a process. Their conflict culminates when Kirk insists, "We're not offering a process" (4.36). If we return to the sessions to reconstruct the purposes and ideas behind the notion of process, we can see why this consensus was more apparent than real and how the terminology that the group chose to use to discuss their plan may have contributed to this misleading consensus.

As early as the first session, and continuing throughout the sessions, these discussions offer evidence of the separately held representations and show why we might expect consensus to break down. Table 1 summarizes the representations held by the four participants during the first session as they discussed the process they hoped to establish, defined the group's goals, and named their purpose. In this session, Liz and Kirk begin by discussing the idea of a process. Already their views suggest a future conflict: Liz sees the process as a well-defined, legal document; Kirk, however, wants to incorporate both legal and social roles (including a discussion of how relationships define responsibilities). In spite of these separately held representations, differences are not confronted and group members proceed to define their goals. In the process of defining goals, differences between LuWanda and Dave also appear. LuWanda, who stresses the need to improve communication between landlords and tenants, contrasts with Dave, who wants to educate the entire community. Again, LuWanda's and Dave's divergent opinions are not openly acknowledged as differences and so resurface when the group names the purpose of their document. For Liz, the goal is to establish expectations; for LuWanda, the goal is to build a relationship. For Kirk, the goal is to clarify responsibilities; for Dave, to create a relationship that involves all of the members of the community with one another. Over the course of the remaining sessions, there is no indication that these four representations shift. By interpreting these representations in light of the personal history and the group's shared history, we may begin to understand why the assumed consensus is misleading.

Liz urges a legal interpretation of process. As a mediator between landlords and tenants, she understands existing processes—the law, the lease, the kinds of decisions judges are likely to make, the kinds of evidence likely to hold up in court. "If there was a process, and we could make it well known," she argues, "even if the landlord didn't participate, he would still know that that process was in place. And that ultimately somewhere down the line, he if, even if he didn't use the process, he would know that he would be taken to task by the process" (3.23). But her concerns are not exclusively with tenants. Because she is so aware of the minefield that landlords and tenants negotiate, she wants "a process that changes the perceptions" of both landlords and tenants (1.24). The process she describes is closely tied to law and based on existing structures. Furthermore, the kind of process she describes can be written down and encoded: landlords need a process for screening tenants (2.24) and for deciding when to evict (1.3); tenants need a process for renting housing (3.14).

Unlike Liz, whose position is clearly articulated and detailed, LuWanda's idea of process is vague. Although she is a tenant and, as a representative for tenants' rights, a vocal representative of the community with experiences to support her own ideas, LuWanda's position often does not get heard. We might suspect that this is so because the others ignore her, but instead it appears that, particularly in the early sessions, she does not speak as frequently as the others (a secondary reason for this may also be that she is late to the first meeting and misses the second one entirely). Perhaps her lack of talk is not surprising—of the two groups being represented, she represents tenants, a group that has historically and socially been thought to have little power. When she does talk, LuWanda calls for "communication" (1.35), wants to develop "the ongoing relationship" between landlords and tenants (1.45), and calls for regular meetings

Table 1. Putting a Moment of Apparent Consensus in Context

Episode	*Liz*	*LuWanda*	*Kirk*	*Dave*
Discussing Topic	A legal process		A global process	
Defining goals	Find remedy	Improve communications		Educate community
		Moment of Apparent Consensus		
Naming Purpose	Establish landlord and tenant expectations	Create understanding and ongoing relationships	Clarify responsibilities	Create ongoing relationships

between landlords and tenants, where individuals can sit down and talk out problems. While this view of simple contact succeeding in solving problems may seem underarticulated and unrealistic, it is a position to which she strongly clings.

Kirk wants a process to articulate responsibilities and to guide community members to responsible action. Unlike Liz, who wants a process that states the rules landlords and tenants must follow, and unlike LuWanda, who wants a process to establish informal communication, Kirk wants a global process to describe how people ought to treat one another. Perhaps because he is skeptical about what any process by itself can achieve, he doesn't see Liz's prescription for more awareness of laws as being useful. Instead he recognizes that "the landlord and tenant would have to buy into that process" and knows that just having a law will not create the sense of buy-in (3.23). Kirk also recognizes that an adversarial relationship already exists between landlords and tenants (1.31) and that this adversarial relationship cannot be overcome simply by notifying individuals of their legal responsibilities. He recognizes flaws in the current system, too (3.27), unlike Liz, who believes that the current system works just fine. All of this, and his own experiences as a landlord, leads him to want to clarify the responsibilities of both landlords and tenants (1.45). This difference in vision comes to a head during the last session (4.25), when Liz and Kirk seem to recognize that they are speaking from very different positions.

So far we have noticed differences in direction (Should the relationship between landlords and tenants be social, communitarian, or legal?) and in depth (To what extent?). Dave, as the full-time, paid president of a community group, has concerns for the landlord, the tenant, and the community around them, and so brings in education but does not advocate for a single position. For Dave, "I think it's gotta be an educational thing" (1.35). He knows that landlords and tenants "tend to develop one viewpoint and fight it to the bitter end" (Self-Interview 1.1) but sees intervention strategies such as role-playing in getting people to see all sides. His primary concern is with organizations, and particularly with "teach-[ing] organizations how to intervene in these squabbles that generally arise" (Self-Interview 1.1). And if landlords and tenants "are educated the same going into the, the agreement—they're both on even ground—or the ground is more even" (1.35). Because of his concern for the entire community, Dave seems to subsume the views of others (like LuWanda, Dave is concerned about communication; like Kirk, he is also concerned about responsibility) to create his own position.

When group members hold disparate representations of a complex problem (based, as we have seen, on multiple factors such as values, experiences, and responsibilities), how can groups achieve agreement? What appears to happen is that individuals may: (1) agree only on terminology, and recognize or fail to recognize that they hold different representations; (2) partially share a particular representation (and recognize, or fail to recognize, that this representation is not shared in its entirety); or (3) believe that their representations are shared in whole. (It is unlikely in our view that representations themselves will actually be shared, although what may be most important here is that the members of the group act as if they were in consensus, that the group's willingness to communicate relies on their willingness to act as if.) If group members walk away from a community discussion believing that their own representation is shared by others in the group, and if it later turns out that this is not the case, it is understandable that group members may end up feeling betrayed, or feeling that community groups do nothing but "talk." It may be then that group members need to see consensus not as a shared understanding of the present situation and desired outcome, but as an agreement to come to action, in spite of the fact that individual representations of the actual problem may not be shared.

Having consensus appear and (because of competing representations) later break down is, we think, a common but critical problem in community groups. This breakdown of consensus leads us to two fundamental questions: What do we mean when we talk about consensus? And what do the members of this community group mean?

Consensus in Action—through Text

Although community members may hold disparate representations, they must still work collectively to accommodate differences. In order to meet their goals, members of this group opted to establish consensus about desirable actions rather than about ideas. In this sense, consensus came about through action—in producing a written text.

For this group, one way to accommodate differences was by developing text conventions that mirrored the collaborative planning strategies of their dialogue. Where the written text typically produced by a community group might state generic problems and solutions, in this project, individuals narrated particular problems that they had encountered or heard about—stories that were typical of the problems found in their community. These narratives, which they called "scenarios," brought to light a variety of problems with landlord-tenant interactions. As a text feature, scenarios structured the final document by providing a place for the group members to present the kinds of problems they saw as typical of their neighborhood. Having raised problems with the scenarios, the group invented another text convention—called "what ifs"—that provided a place in the text to suggest possible solutions. "What ifs" are questions that are physically appended to scenarios and which expand the possible ways of resolving the problem that has been identified. These "what ifs" raised questions about what could have been done and suggested alternatives for handling the situation. Both the scenarios and the "what ifs" provided a structure for the final text, but they went beyond merely structuring the text because they also served social and personal functions.

Scenarios allow individuals to feel personally engaged and to bring concrete experiences to the negotiation. By telling stories and sharing information about themselves and their experiences, they also allow the group members to share their values, beliefs, and concerns about the community and its problems. This, in turn, allows individuals to have their perceptions reinforced by the group. When Dave describes a problem he faces ("we have tenants in [our neighborhood] that know how to use the system, how to stay in their apartment for 6 months [without paying rent]" (1.36)), it allows Kirk to describe a similar experience ("the tenant would come to me and say, [my employer] got screwed up . . . [and so I can't pay my rent] and the next month [the date that I receive my rental payment] will be later" (1.39)). While these stories may be seen as grounding the discussion, they also ground the individual, by helping the others see that personal experiences shape the speaker's understanding of the community and its problems. Furthermore, by sharing their experiences, the group can begin to create a shared understanding of the community in which they live.

Scenarios bear the stamp of the individual writer's goals. Dave's goal, for example, of having landlords and tenants recognize that their actions will also affect the community, is played out in his scenario. When neither the landlord nor the tenant meets the other's expectations, his scenario reveals, the entire community suffers. As the president of a community group, we can expect that his concern is with long-term community issues, and this is reflected in his scenario. Likewise, Liz's concerns as a mediator are reflected in her scenario, which addresses how people ought to act in particular situations. Similarly, LuWanda's concern for communication and Kirk's concern for responsibility are revealed in their respective scenarios.

Scenarios serve social goals too by making it possible for group members to share particular experiences and to think through ways each of them might approach the problem. In doing this, the knowledge, beliefs, and values of each individual are applied to a shared problem. Group members can, in this process, hear other perspectives and come to understand other members of their community. Additionally, group members may have their own ideas validated by others in the group. Relationships within the group are built and strengthened in the process of sharing problems and working together to find solutions.

Dissensus is acknowledged and invited out into the open with scenarios. After Dave reads his own scenario, for example, Kirk wonders aloud when the landlord ought to have evicted the tenant who has stopped paying rent. In the ensuing discussion, it is clear that the members of the group have very different opinions. But asking the question allows the group to consider a number of possibilities and to reflect on its own practices—the same sort of active reflection that they hope to create in their readers. Thus, scenarios, in fact, not only acknowledge dissensus but also allow dissensus to surface for the purpose of inquiry.

And scenarios are able to accommodate that dissensus in the final text. Scenarios let group members reflect on the conflicts and potential conflicts they are describing; rather than trying to resolve and come to consensus about these points of disagreement, using this convention lets them incorporate their own lack of agreement into the document. While they see scenarios as an effective strategy for reflecting the differences that exist between landlords and tenants, the notion of scenarios really emerged as a way to respond to the dissensus that existed. They had given themselves the goal of describing the relationship that ought to exist between landlords and tenants, but they were still divided over what kind of relationship ought to exist, and in particular over whether their job as representatives of the community was to prescribe a code of behavior or to set options for behavior. As described above, Liz advocated setting standards, but Kirk wanted to help members of the community think through their options. The scenario and "what if" conventions let them accommodate these apparently mutually exclusive goals. Rather than glossing over the disagreements that actually existed within their group, they ultimately decided to construct a text that not only reflected, but even capitalized on, those differences.

In addition to accommodating difference, this text convention evolved through a process of negotiation. Tracking this evolution reveals how members constructed a strategy for dealing with differences. Over the course of the sessions, the participants negotiated what the word "scenario" meant. Liz, like a negotiator working with particular instances, wants to know if as part of their discussion they will be given a scenario to try to resolve (1.5); here she seems to be expecting a particular, existing conflict. But for this group scenarios are not always about situations that have already occurred. For Kirk, scenarios seem to represent a way to open up general problems and to see them from a number of different perspectives. He talks about "scenario building" (1.27), recognizes that "there are other scenarios" (1.26), and wants the group to "finish out that scenario" (1.27). For a time, it seems doubtful that scenarios will really be of much use to the group because of this tension about what scenarios ought to do. But gradually the idea is modified.

The features change over time in order to accommodate differences. At first, scenarios appear as models and examples for how individuals ought to behave. But later, Dave introduces the text convention of "what ifs" as a way to think through a problem: "If we do a scenario thing, we could say, now 'what if' you did this, and 'what if' you did that?" (2.28)—a move that also seems to accommodate Kirk's desire to provide options. Scenarios are also seen as establishing certain "parameters" for acceptable action. As Kirk suggests, "the parameters of this scenario are sort of clear: you've got a bad tenant" (2.41). And this feature of scenarios seems to some extent to meet Liz's needs. The advantage of the scenario is that it can evolve to accommodate the goals of all of these writers. This cumulative development of the concept of scenarios makes them a flexible tool in resolving the group's tension about what they ought to do. We are not claiming that scenarios in and of themselves are the necessary solution; in fact, we would guess that any number of textual features could have served a similar function. Instead, what seems essential here is that the group members knew how to use a text to represent strongly felt personal problems while at the same time moving the group toward social action.

Balance of Power

But what determines what is actually included in these scenarios and in the final, completed text? We have already argued that individuals come to the discussions with strong ideas about the problems facing their community and how to solve them. We have also argued that often differences between individuals are not resolved in the process of coming to consensus. Yet it seems clear that there must be some way to reconcile these differences (after all, a final text does exist and the members of this group have come to some sense of resolution in agreeing on its contents). If it is not consensus alone that allows these individuals to produce the final text, then what is it? We might assume that strong social norms (expertise, experience, and education, for example) are at work, deciding what voices are heard. For example, we might expect that the landlords (Dave and Kirk), as owners of property, might be given more attention. Or we might expect that leaders in the community (Liz as a mediator, Dave as a president of a community group) would have more prestige and would claim more authority. Or we might expect that the men (Kirk and Dave) would be given (or would take) more time to talk than the women (LuWanda and Liz). And we would probably also expect experience and education and a number of other factors to play some role in how these different individuals would get their messages across. For instance, all of these points might lead us to believe that as the discussion continued, LuWanda's voice would be heard less and less, and that her ideas would be unlikely candidates for inclusion in the final text. But is this what actually happens?

When we count conversational turns (see table 2), a pattern of interaction emerges that seems to hold true for at least some of our expectations. LuWanda, as expected, has the fewest number of conversational turns. Liz, the experienced mediator, has the most (with over three times more than LuWanda). Dave and Kirk are roughly equivalent to one another, with more than 850 conversational turns each (almost twice as many as LuWanda's 480).

Table 2. Total Conversational Turns for Each Participant

Liz	1487
Dave	893
Kirk	868
LuWanda	480

But such a chart, which merely reflects totals, doesn't adequately describe the interactions occurring within particular sessions. In order to do this, we need to examine what these numbers represent.

Table 3, which maps out differences in the number of conversational turns in each of the sessions, reveals a slightly different pattern. It reveals, not surprisingly, that everyone tended to talk more during later sessions than during earlier ones. It also shows Liz, the mediator, consistently talking more than anyone else. What does seem surprising, however, is that LuWanda, who talked least frequently during the first three sessions and who walked into the group with the least "power" (in the sense that we have defined above), was at least as involved as either Kirk or Dave during the final session.

Table 3. Conversational Turns for Each Participant Per Session

	1st Session	2nd Session	3rd Session	4th Session
Liz	288	335	316	548
Dave	124	307	174	288
Kirk	134	230	222	282
LuWanda	95	—	91	294

LuWanda, because she seems to break from our expectations, offers an interesting place to begin an investigation. As she describes her motivation for participating, she says, "What I wanna accomplish in this meeting is to be, to be heard" (Self-Interview 1.2). More pragmatically, "I wanna know how to go about discussing, getting these things done, coming to some sort of agreement to get this situation resolved [describing a problem she is having with her own landlord], or will it ever be?" (Self-Interview 2.7–8). LuWanda, who clearly seems motivated to participate, does not seem to know how to reach her goals, yet she is successful in ensuring that the final text does include a discussion of "communication and responsibilities." Given LuWanda's sense of her limited authority, how is she able to have her voice heard in the discussion and to have her ideas included in the final text? (It is important to note here that we are not claiming that the other members of the group found her position gained acceptance simply because she talked more, nor do we assume that the act of talking about a position grants it greater validity or authority with a group.) Still, if our initial hypotheses are accurate concerning LuWanda's lack of overt involvement early on (which we see as being linked to her own sense of authority, rather than her personality), then it seems likely that either something about the group changed, or LuWanda herself did something differently to position herself to be involved in the discussion. What did she do to enable her "voice" (in the double sense of physical voice and her ideological voice) to be heard?

As we saw in the analysis of misleading consensus, these writers do not concur on how to represent this problem. By personal experience and by design they bring alternative perspectives and different—even competing—interests to the table. They are committed to positions that reflect personal histories and the people they represent. Moreover, the design of Argue validated diversity—the positions of landlord and tenant, the professional, personal, and community perspectives, and cultural differences of black and white.

Consensus as the Construction around Conflict

So what does it mean to say they achieved consensus? This argument process does not fit the model of debate in which a contest between preestablished claims is resolved on the basis of logic or in which through persuasive appeals to evidence or emotion one set of propositions emerges the victor. This is not a democratic consensus in which consensus forms around the magnet of a majority will. Nor is this like the consensus achieved in the negotiation of a dispute or contract in which participants come with fixed goals, but willing to barter minor outcomes to achieve major ones, although there are important parallels. For instance, one of the key strategies developed in the Harvard Negotiation Project is apparent here as participants try to shift the ground of negotiation away from particular outcomes to larger values and grounds of agreement (Fisher and Ury).

In contrast to the models of debate, democracy, or arbitration, the consensus we see here is achieved not around a set of winning claims, a dominant perspective, or a distribution of benefits, but through a joint literate action. The group may not agree on the problem or even on the best, wisest, most effective ethical response, but they were able construct a shared action they could take in text.

Looking at this event as a collaborative literate action in which people are constructing a meaning helps explain how conflict became at times a generative, inventive force, rather that a mere contest of entrenched positions. First, the design of the collaborative planning process gave people new roles (beyond that of spokespersons) as collaborative partners dedicated to the construction of something concrete—a text that would help their community deal with a problem. People were positioned from the beginning not only as problem solvers but as writers and collaborators, whose job included quite literally helping the "opposition" articulate their vision.

Secondly, this literate action became a site for consensus because it allowed a multi-vocal representation of the problem. The negotiation theory which we are bringing to this analysis argues that when writers are pushed to the negotiation of conflict they must attend to a circle of voices that advise, suggest, cajole, trouble, persuade and generally attempt to shape meaning in their own image (Flower, *The Construction*). Literate action, here in the forms of scenarios and "what ifs," offered a way to incorporate a variety of voices, not as opposing claims but as conditionalized ways of responding to a complex situation. In short they tried to construct a meaning that supported action in the face of diverse facets of this problem.

Finally, this literate construction supported consensus because it was happy to cross boundaries and violate the expectations of conventional texts. The final document, like other texts emerging from the CLC, was a hybrid text. On one level, it was a published document, focused on four critical problems, motivated by policy issues of how to interpret the complex causes of landlord and tenant conflicts. But at the same time, it was designed to speak to these problems, not in the conventional language of policy or analysis, but out of the experience of everyday people in the community, telling stories. Yet, unlike the landlords' "horror stories"—one-sided accounts of commiseration—the scenarios and "what ifs" are grassroots policy statements designed to reflect the competing perspectives that underlie the conflict. And as a grassroots approach to policy, the document did not present itself as a traditional policy statement in which analysis was shrinkwrapped into a set of more abstract recommendations. Instead, the group decided to design the document as the basis for a community discussion they held that spring, using the scenarios as a way to invite other organizations into the construction of a larger discussion and text.

Landlords and Tenants is an example of a mixed genre, an eclectic, ad hoc text design invented to serve a rhetorical purpose. However, as a hybrid text it negotiates a even more important set of conflicts when it crosses the boundaries of powerful, socially significant discourses—discourses that rarely occupy the same space. As the draft evolves, the legal, procedural language of certainty rubs shoulders with story telling, and warm, African American statements of conviction and adjuration.

Hybrid discourses like these are an important alternative to single-voiced documents not only because they invite different perspectives, but because they allow a larger community to hear themselves speaking in the document in the language they use, in a discourse that empowers rather than marginalizes them. Such documents allow more people to feel they too can stand in the discussion—they are being spoken to and invited to speak back.

Conflict and Transformation

Consensus in action and hybrid texts let writers negotiate conflict by not only recognizing difference, but embracing it. They navigate among conflict and constraint looking for the best path, trying to construct a meaning that listens to multiple voices and preserves as many values as possible. However, at times negotiation in the group dealt with conflict head on, when differences were put on the table as explicit disagreements. In these circumstances, it would be reasonable to expect the sort of win-lose scenarios both Kirk and Dan described, contests in which social norms and prestige discourse dominate the construction of meaning. In the face of such expectations, the process of knowledge transformation we did observe becomes doubly interesting.

Clearly other strategies (beyond social norms) are being invoked.[6] What are these strategies? And how are they used? Here we will focus on LuWanda, who employs a number of strategies to bring her views into wider circulation, but we will also show that the others used strategies to reshape their ideas in order to make them more appealing to members of the group as well. Strategies for transforming ideas include linking new ideas with accepted ideas, repetition, and adapting terminology. Here we define and discuss briefly the rationale these individuals may have used in opting for their use.

One way of transforming ideas is by linking weaker ideas with stronger ones. While the group's purpose is initially framed as an investigation into the rights and responsibilities of landlords and tenants, the group has already been thinking along these lines. In fact, Kirk's interest in rights and responsibilities began long before this group discussion: "I have been a landlord for many, many years and am constantly trying to figure out them, this rights and responsibilities idea" (1.7). And this linking of rights and responsibilities seems acceptable to each of them: Liz agrees when she insists that "you can't speak of, of rights alone, you have to add rights and responsibilities" (1.8). LuWanda also accepts the linking of rights and responsibilities, although she sees problems resulting from people's inability to communicate with one another. Early on she positions herself: "I can understand both sides and I have arguments on either side, so my main purpose is to see if there can be communication between the landlords and the tenants maybe once a month, that they could talk about the issues that need to be addressed" (1.11). LuWanda, whether knowingly or not, uses an idea that has already gained acceptance within the group ("rights and responsibilities") to launch her primary concern: communication. In her mind, balancing rights and responsibilities is not possible without communication. And this link begins to hold for other members of the group.

Even after the group seems willing to accept communication as one of the major concerns of the group (Liz agrees as early as the first session (1.21) that communication should be one of their key points), LuWanda returns again and again to the idea. During the first session, she introduces and reinforces, repeating it six times (considerable when we take into account that she has a total of 95 turns). (Since LuWanda isn't there during the second session, it's important that she's already gotten her main point in during the first session.) While she continues to reinforce her idea during later sessions, it is this first session that sets her position and introduces others to it. Repetition, then, is another strategy that LuWanda uses to reinforce her ideas.

Rather than simply advancing the idea of communication, however, LuWanda starts talking about "communication and responsibilities." Her use of the phrase seems strategic too, in that communication precedes responsibilities and replaces "rights." Although there is little sense that she is aware of her own strategies, this is the final phase in making her ideas acceptable: first, she links her own ideas to those that have already been accepted. Next, she repeats her ideas to reinforce the group's acceptance. Finally, she transposes her term ("communication") so that it replaces another ("rights") but also precedes "responsibilities." In this process, the discussion of "rights" drops out and the group continues discussing communication and responsibilities. The other group members may not initially intend for this to happen. Dave sees communication as "one of the responsibilities, I mean, it's probably one of the key responsibilities" (11), but he clearly sees it as being subordinate to "responsibilities" alone. Nevertheless, the linking of communication and responsibilities seems to stick, with Dave arguing that "the purpose is communication and responsibility" (2.54), and Kirk supporting with, "And basically it seems that most of these conflicts are lessened with good communication and people taking responsibility" (2.55). Liz, when asked about the key point to be made in the document, responds with "how it can create a better working relationship between the tenant and the landlord you know, by communicating and, and by communicating the responsibilities of each individual a better relationship can be created" (2.55). Her stress is on the responsibilities more than on communication, but she seems willing to use the ideas that others have advanced in order to ensure that the main point is stressed. (Liz seems primarily concerned about how these ideas will be received by the audience—when Lorraine asks her why they should discuss both communication and responsibility, she argues, "that's a good idea because that's really a positive way of doing it" (2.57). This is formalized when Dave argues, "it is our intention to discuss communication and responsibility between tenants and landlords" (2.61), and Liz states, "I think we all agree that communication and responsibility are two key elements in resolving or dealing with tenant landlord conflicts" (2.62). Listen to Liz try to formalize the group's agreement: "We are, as individuals, going to take a specific conflict and try to get a clear outline or . . . the cause and effect of the conflict and a positive outcome. Dealing with communication and responsibility" (2.66). By linking communications with responsibilities, LuWanda begins a strategic process that ensures "communication" (and not just "rights") is taken into account.

While the terminology being used seems important to how quickly the ideas will be accepted, the individuals seem more concerned about finding ways to form relationships between ideas that have already been accepted by the group and the particular ideas they are interested in. The whole group, for instance, is interested in "responsibility." When individuals talk about responsibility, they know they will find an audience. What seems less important to them is that their ideas about what constitutes responsibility are divergent. When Kirk talks about responsibility, for example, the implication is "social responsibility" (1.41), particularly when he describes "people [who] try to relinquish more and more responsibility for what happens in their lives" (2.3) and argues that "the best we can do is outline responsibility" (2.31). In contrast, when LuWanda talks about responsibility, she is really referring to the responsibility to communicate. And Liz's sense of responsibility is the ability and willingness to engage in a legally defined relationship. Yet in spite of their different conceptions of what responsibility means, they all rely on the same terminology to present their ideas to the group. During the course of their discussions these differences emerge, yet each continues to use "responsibility" to define the concepts each is most concerned about.

Furthermore, neither strategies for transformation nor individual positions seem to change as the result of having other members of the group discount certain ideas. Instead, what seems to occur is that individuals whose ideas are not accepted by the group (particularly when those ideas are strongly felt) continue to bring their own ideas up as reasonable alternatives to the ideas that are offered by other members of the group. Liz, for example, is unable to get the others to embrace her interest in legal responsibilities, but this does not stop her from continued discussion. During her self-interview, she shows her own awareness of this phenomenon: "I really do have my own agenda when it comes to this memorandum of understanding and I'm sure that it differs very much from those of people that I have been sharing this group with" (2.1). For Liz, responsibilities are more important than communication (3.17) because the legal relationships define what kinds of communication are possible. LuWanda takes a similar approach. For her, the stress is on "communication"—in fact, her scenario reflects only her concern for communication, in spite of her agreement that the group's focus is on "communication and responsibilities" (4.26). LuWanda argues: "You can't find out about your responsibilities unless you communicate. That's how, that's the basis around responsibilities, is communicating to find out what you do agree on and what you're gonna do, you have to talk, you have to" (4.72). When Lorraine asks about the key point of the scenarios, and prompts with "the key to resolving these problems is . . ." Kirk and LuWanda both reply "communication" (4.70), but Liz responds, "Communication and responsibilities." Later she urges, "I think we need to talk a little bit more about responsibilities" (4.72). Both Liz and LuWanda claim they are willing to accept the group's decision to focus on communication and responsibilities, but each continues to act in her own interest.

Whether knowingly or unknowingly, members of this group used a variety of strategies in order to transform how their ideas were received by other members of the group. These rhetorical strategies seem to have been successful in ensuring that voices are heard, sometimes at the cost of sustaining the conflict that community members are trying to minimize. In community groups, the desire for the group to be in a state of consensus often runs up against the desire of the individual to have his or her voice included in the final text. As a result, conflicts are not brought to the surface (to discuss them directly would be to rend the fabric of consensus) and are left to emerge again and again. For this group, reaching consensus means reaching a point where conflicts are hidden (but certainly not gone) in order to create a solid foundation for moving to action.

What Conclusions Can We Draw?

In this situation, three key features seem to have affected the final text. First, in this situation the goal was not to produce a text that would articulate their shared values. Rather, the goal was to move members of a community group towards action. Second, although this group was composed of a number of individuals who represented differing (and often opposing) positions, the group recognized that they were accountable to one another. That is, members of this group recognized that failing to include competing ideas would mean that others in the community might not be able to find their position in the document, and that this would work against their desire to move the group towards action. Third and finally, not only were the members of the group accountable to one another, but they were personally invested in the outcome of the group. They were not trying to tell others what to do, but were instead trying to find some practices that the group could live with.

Under these conditions, invention seems to be not only about finding the available means of persuasion that will lead the readers to agreement about a particular conclusion. In fact, invention here seems to be, at least in part, about finding a way to get ideas across so that the tension of the conflict is actually made explicit. Here the text is not a final product, a culmination of the group's thinking, but is simply part of an ongoing discussion, where conflicts are externalized in writing and made part of the group's shared knowledge. A rhetoric of pluralism requires a new understanding of invention—what is required are tools for invention that will recognize that consensus, even while it may support the group's willingness to work together, is often ephemeral. Here invention needs to support dissensus while ensuring that the focus of the group remains on action. Similarly, here invention's foremost goal may not be that all arguments are systematically considered, but that conflict can be recognized and represented. In an attempt to focus on the long-term goal of bringing a group together to discuss a problem, share ideas, articulate conflicts, and create a plan for action, writing can play an integral role. Perhaps part of learning to invent should be learning ways to articulate conflict in text as a basis for the collaborative construction of a more complex meaning.

Notes

Office of Educational Research and Improvement disclaimer: The research reported herein was supported by the National Center for the Study of Writing and Literacy under the Educational Research and Development Center Program (grant number R117G10036) as administered by the Office of Educational Research and Improvement, U.S. Department of Education. The findings and opinions expressed herein do not reflect the position or policies of the Office of Educational Research and Improvement or the U.S. Department of Education. We want to thank Lorraine Higgins, who not only played a major role in the conduct of this study, but in our thinking and interpretation through her own work and our collaborations with her.

1. For a more theoretical discussion of community literacy and the aims of the CLC see Peck, Flower, and Higgins (1995). Long (1994) and Flower (1996) document the work with teens and the mentors' own attempts to interpret this intercultural collaboration. Flower, Wallace, Norris, and Burnett (1994) offer background on collaborative planning in educational contexts.

2. An analysis of how the educationally based practice of collaborative planning was adapted to this community context can be found in Higgins, Flower, and Deems (in press). This attempt to bridge discourses builds on Higgins's (1992) theoretical discussion of argument construction that guided her study of how women, returning to an urban community college, negotiated conflicting styles of academic and community argument.

3. The names of the community residents who participated in this project have been replaced with pseudonyms throughout this paper.

4. Julia Deems joined the project (as a NCSWL Research Assistant) as the analysis of data began, coauthoring the interpretation presented here. We wish to thank Elenore Long, who managed the collection and organization of the data.

5. Here the first number ("1") indicates the quote comes from the third session, and the second number ("45") indicates that it comes from the fourteenth page of the transcripts from the third session. We use this convention throughout the rest of this document.

6. Although we describe this as a strategic process, when participants were asked about their own strategies, we received a variety of responses. Kirk, for example, was entirely aware of his tendency to play devil's advocate and how his vocabulary choices were predicated on how he wanted his ideas to be received. LuWanda, on the other hand, did not talk about her strategies. When questioned about whether she did in fact have strategies, LuWanda did not show evidence that she knew she was using them. Nonetheless, LuWanda was using strategies. We would argue, therefore, that people are not necessarily aware of how they carry on the process of transformation, but that transformation occurs in any case.

Works Cited

Bruffee, Kenneth A. "Collaborative Learning and the 'Conversation of Mankind.'" *College English* 46 (1984): 635–52.

Donnellon, Anne, Barbara Gray, and Michel G. Bougon. "Communication, Meaning, and Organized Action." *Administrative Science Quarterly* 31 (1986): 43–55.

Enos, R. L., and J. M. Lauer. "The Meaning of Heuristic in Aristotle's *Rhetoric* and Its Implications for Contemporary Rhetorical Theory." *A Rhetoric of Doing: Essays on Written Discourse in Honor of James L. Kinneavy*. Ed. Stephen P. Witte, Neil Nakadate, and Roger D. Cherry. Carbondale: Southern Illinois UP, 1992. 79–87.

Fisher, Roger, and William Ury. *Getting to Yes: Negotiating Agreement Without Giving In.* New York: Penguin, 1983.

Flower, Linda. *The Construction of Negotiated Meaning: A Social Cognitive Theory of Writing.* Carbondale: Southern Illinois UP, 1994.

———. "Literate Action." *Composition in the Twenty-first Century: Crisis and Change.* Ed. Lynne Z. Bloom, Donald A. Daiker, and Edward M. White. Carbondale: Southern Illinois UP, 1996. 249–60.

Flower, Linda, David L. Wallace, Linda Norris, and Rebecca E. Burnett, eds. *Making Thinking Visible: Writing, Collaborative Planning, and Classroom Inquiry.* Urbana: NCTE, 1994.

Habermas, Jurgen. *The Theory of Communicative Action.* Vol. 2. Boston: Beacon, 1984.

Halliday, Michael Alexander Kirkwood, and Ruqaiya Hasan. *Cohesion in English.* London: Longman, 1976.

Higgins, Lorraine. "Argument as Construction: A Framework and Method." Diss. Carnegie Mellon U, 1992.

Higgins, Lorraine, Linda Flower, and Julia Deems. "Collaboration for Community Action: Landlords and Tenants." *Collaboration in Professional and Technical Communication: Research Perspectives*. Ed. Rebecca E. Burnett and Ann Hill Duin. Mahwah: Erlbaum, in press.

Latour, Bruno, and Steve Woolgar. *Laboratory Life: The Construction of Scientific Facts.* Princeton: Princeton UP, 1986.

Long, Elenore A. "The Rhetoric of Literate Social Action: Mentors Negotiating Intercultural Images of Literacy." Diss. Carnegie Mellon U, 1994.

Miller, Carolyn. "Rhetoric and Community: The Problem of the One and the Many." *Defining the New Rhetorics.* Ed. Theresa Enos and Stuart C. Brown. Newbury Park: Sage, 1993. 79–94.

Peck, Wayne C., Linda Flower, and Lorraine Higgins. "Community Literacy." *College Composition and Communication* 46 (1995): 199–222.

Perelman, Chaim, and L. Olbrechts-Tyteca. *The New Rhetoric: A Treatise on Argumentation.* Trans. John Wilkinson and Purcell Weaver. Notre Dame: U of Notre Dame P, 1969.

Rescher, Nicholas. *Pluralism: Against The Demand for Consensus.* Oxford: Clarendon, 1995.

Stubbs, Michael. *Discourse Analysis: The Sociolinguistic Analysis of Natural Language.* Chicago: U of Chicago P, 1983.

Trimbur, John. "Consensus and Difference in Collaborative Learning." *College English* 51 (1989): 602–16.

Invention, Critical Thinking, and the Analysis of Political Rhetoric

DONALD LAZERE

During the past decade, America's culture wars have been fueled by the introduction of political subject matter into college composition, as well as literature, cultural studies, and other humanities courses. As scholars of rhetoric, we might well be surprised to find such acrimony both outside and within our profession, since the study of political discourse has been at the center not only of the Western rhetorical tradition but also mainstream educational reform since the 1970s. In 1975, NCTE passed the following resolution:

> Resolved, that the National Council of Teachers of English support the efforts of English and related subjects to train students in a new literacy encompassing not only the decoding of print but the critical reading, listening, viewing, and thinking skills necessary to enable students to cope with the sophisticated persuasion techniques found in political statements, advertising, entertainment, and news.

Likewise, the Rockefeller Foundation's Commission on the Humanities in 1980 emphasized that "the humanities lead beyond 'functional' literacy and basic skills to critical judgment and discrimination, enabling citizens to view political issues from an informed perspective" (*The Humanities* 12). The report continued:

> The humanities bring to life the ideal of cultural pluralism by expanding the number of perspectives from which questions of value may be viewed, by enlarging young people's social and historical consciousness, and by activating an imaginative critical spirit. . . .
>
> The entire secondary school curriculum should emphasize the close relationship between writing and critical thinking. . . . English courses need to emphasize the connections between expression, logic, and the critical use of textual and historical evidence. (30, 44)

Much of the controversy, of course, has not involved the value of teaching political literacy as much as the very real danger of courses being turned into an indoctrination to the instructor's particular ideology—in recent years usually some variety of leftism—or, at best, into classes in political science rather than composition and rhetoric. (Probably the best-known expression of concern over this danger from within the English profession was Maxine Hairston's 1992 *CCC* article "Diversity, Ideology, and Teaching Writing.") In my opinion, this controversy has not resulted from an excess of the study of political rhetoric in college education but from a deficiency of it in recent decades, during which the dominant paradigms of both rhetorical theory and composition instruction have failed to provide useful models either for systematic scholarly study of the current practices of political discourse or for instructing students to be critics of and participants in that discourse. When the heated political battles of the sixties inevitably entered the classroom, the profession had no established guidelines for ethics in the practice or teaching of political controversy. Nor was much cognizance taken, either by the profession at large or by many of the teachers who introduced political material, of the reality that today's entering college students are, for the most part, woefully lacking in elementary knowledge of political history, terminology, and ideology. Because they are unprepared to deal with any political subject matter, they are indeed vulnerable to manipulation by partisan teachers—though most are probably more bewildered than converted by professorial advocacy. However, attempts like Hairston's to sanitize composition instruction against any possibility of political coercion by teachers have simply evaded the issue of students' political naiveté and led toward the abdication of our legitimate responsibilities as scholars to maintain the classical study of civic discourse and as teachers to prepare students to take an active part in the deliberations of the polis. In fairness to Hairston and other critics who have assumed that it is basic or first-term writing courses that are being politicized at the expense of more appropriate instruction at that level, I agree, to the extent that this has in fact taken place. My emphasis on political rhetoric is not keyed to this level but to advanced writing courses in argumentation and research, which, unfortunately, have received relatively little attention in recent composition theory.

What is called for, then, rather than depoliticizing composition studies, is developing—under the rubric of rhetorical invention—ethical, practical methods for expressly studying and teaching the very topic of political controversies, including all the subjective biases, partisan polemics, and propaganda that color them. This article will explain the very tentative, open-ended approach to these problems that I have devised over the past decade. After placing this approach in the context of current theories of rhetorical invention, I will trace its grounding in the disciplines of critical thinking and general semantics; then I will review its applications in my previous scholarly work and course outlines, culminating in four heuristic sets of guidelines for the critical evaluation and practice of political arguments.

Paradigms for Political Analysis

Several of my recent publications have applied Gerald Graff's notion of "teaching the conflicts" to the analysis of contemporary American political rhetoric; this has been a theoretical attempt to develop a taxonomy of patterns of political argumentation, as well as a practical attempt to provide students with interpretive heuristics for understanding and evaluating the arguments they encounter every day in media of news, opinion, and entertainment, in peer discussion, and in mass-mediated electoral and legislative politics per se. This approach differs from models of invention deriving from discipline-specific academic studies in the social sciences or communications, although it may well be useful in application to such studies; indeed, its disciplinary base is argumentative rhetoric itself, by way of schemas derived from theories in critical thinking, general semantics, and cultural studies, within a broadly Aristotelian tradition. These schemas are also applications of James Berlin's social-epistemic rhetoric: "the study and critique of signifying practices in their relation to subject formation within the framework of economic, social, and political conditions," with particular emphasis on the self-reflexive examination of ideological subjectivity in rhetoric and rhetoricians (Berlin 77).

My project, then, has been to devise a distinctly rhetorical framework within which students (and citizens in general) can learn to understand

and participate in political controversies, guided but not indoctrinated by teachers or scholars. In principle, this framework would approach ideological conflicts, not through the monologic perspective of any teacher's or scholar's own beliefs, but through dialogically enabling students to understand and evaluate a full range of opposing ideological perspectives, their points of opposition, and the predictably partisan patterns and biases of their rhetoric—that is, not only what lines of argument partisans of different viewpoints make, but each one's distinctive means of persuasion and degree of adherence to principles of ethical argumentation.

Does my project fall clearly within this book's topic of invention as heuristics or discovery? Strictly speaking, I suppose not, if these terms are limited to mean the production of writers' own, original arguments, rather than analysis and evaluation of arguments in sources. Some rhetoricians would categorize my approach as hermeneutics, in the sense of providing tools for interpreting and evaluating the texts of political and cultural discourse, but I do not find this a crucial distinction. Obviously, readers can benefit from discovery tools for identifying and critically evaluating lines of argument in sources (e.g., political speeches, news and opinion articles, talk radio, as well as scholarly works), then go on to incorporate their analyses in producing a variety of their own texts in a variety of rhetorical situations. For example, I have assigned students in my argumentative writing courses to apply the rhetorical analyses in their term papers in class discussions, speeches and debates, articles for student or professional publications, and letters to legislators.

For my purposes here, however, I am more concerned with the process of rhetorical analysis of sources than with its ultimate applications, my reason being that the focus of composition studies in recent decades has privileged the production of students' own texts to the neglect of the rigorous process of learning to evaluate source material that is a prerequisite to self-expression in most academic studies. The postmodernist turn in composition theory, with its emphasis on multicultural voices, "stories," and nonacademic writing, has been valuable toward its own goals, but it has frequently ignored, or even denigrated, the need for writing instruction as a liaison to academic discourse—a central aim in Mina Shaughnessy's *Errors and Expectations,* which has unjustly fallen into disfavor among postmodernists. Of course, this should not be an either-or choice, but it is foolhardy to expect students to express themselves or argue individually or in groups about subjects like politics and political rhetoric, of which they have virtually no prior knowledge. In other democracies today, and in earlier periods of American history, students have gained adequate grounding in civic education prior to college, so writing instruction at that level could build on such grounding; however, most students in American colleges today—from nearly all social and ethnic backgrounds—get such shallow civic education in high school that remedial instruction in reading, writing, and reasoning about politics is needed as much as in reading, writing, and reasoning in general. The paradigm of writing instruction as a guide to production of students' texts based primarily on their own experiences and in communal settings, then, has served by default to neglect the vast remedial effort needed in the cognitive or epistemic realm toward civic literacy. In the following section, I survey some of the cognitive-epistemic models that I have found most helpful as guides to discovery or invention in this remedial effort.

Critical-Thinking and Semantic Models for Rhetorical Invention

As previously noted, the approach to scholarship and teaching that I am sketching out here has evolved in keeping with calls by educational leaders since the late seventies for increasing emphasis in secondary and college curricula on developing critical reading, writing, and reasoning skills in application to public discourse. In 1980, Chancellor Glenn Dumke announced the requirement of formal instruction in critical thinking throughout the nineteen California State University campuses, serving some three hundred thousand students. The announcement read:

> Instruction in critical thinking is to be designed to achieve an understanding of the relationship of language to logic, which should lead to the ability to analyze, criticize, and advocate ideas, to reason inductively and deductively, and to reach factual or judgmental conclusions based on sound inferences drawn from unambiguous statements of knowledge or belief. The minimal competence to be expected at the successful conclusion of instruction in critical thinking should be the ability to distinguish fact from judgment, belief from knowledge, and skills in elementary inductive and deductive processes, including an understanding of the formal and informal fallacies of language and thought.

Similar requirements were soon adopted by community colleges and secondary schools throughout California and elsewhere. Here is the list of "Basic Critical Thinking Skills" in the California State Department of Education's Model Curriculum for Grades 8–12 in 1984.

1. Identify similarities and differences

The ability to identify similarities and differences among two or more objects, living things, ideas, events, or situations at the same or different points in time. Implies the ability to organize information into defined categories.

2. Identify central issues or problems

The ability to identify the main idea or point of a passage, argument, or political cartoon, for example. At the higher levels, students are expected to identify central issues in complex political arguments. Implies ability to identify major components of an argument, such as reasons and conclusions.

3. Distinguish fact from opinion

The ability to determine the difference between observation and inference.

4. Recognize stereotypes and clichés

The ability to identify fixed or conventional notions about a person, group, or idea.

5. Recognize bias, emotional factors, propaganda, and semantic slanting

The ability to identify partialities and prejudices in written and graphic materials. Includes the ability to determine credibility of sources (gauge reliability, expertise, and objectivity).

6. Recognize different value orientations and different ideologies

The ability to recognize different value orientations and ideologies. Values which form the common core of American citizenship . . . will receive primary emphasis here.

7. Determine which information is relevant

The ability to make distinctions between verifiable and unverifiable, relevant and nonrelevant, and essential and incidental information.

8. Recognize the adequacy of data

The ability to decide whether the information provided is sufficient in terms of quality and quantity to justify a conclusion, decision, generalization, or plausible hypothesis.

9. Check consistency

The ability to determine whether given statements or symbols are consistent. For example, the ability to determine whether the different points or issues in a political argument have logical connections or agree with the central issue.

10. Formulate appropriate questions

The ability to formulate appropriate and thought-provoking questions that will lead to a deeper and clearer understanding of the issues at hand.

11. Predict probable consequences

The ability to predict probable consequences of an event or series of events.

12. Identify unstated assumptions

The ability to identify what is taken for granted, though not explicitly stated, in an argument.

Scholars in developmental psychology, sociolinguistics, and composition theory have supplemented such criteria with other skills of analysis and synthesis that distinguish advanced stages in reading, writing, and reasoning (sometimes termed "higher order thinking"). These include the abilities to reason back and forth between the concrete and the abstract, the personal and the impersonal, the literal and the hypothetical or figurative, and between the past, present, and future; also, the abilities to retain and apply material previously studied and to sustain an extended line of argument in reading, writing, and speaking, incorporating recursive and cumulative thinking (the abilities to refer back to previously covered material and to build on that material in developing stages in an argument).

Some scholars make a further distinction, between critical thinking skills, related formally or informally to traditional logic, and dispositions that foster or impede critical thinking within the broader context of psychological, cultural, social, and political influences. Dispositions that foster critical thinking include facility in perceiving irony, ambiguity, and multiplicity of meanings or points of view; the development of openmindedness, autonomous thought, and reciprocity (Piaget's term for ability to empathize with other individuals, social groups, nationalities, ideologies, etc.). Dispositions that act as impediments to critical thinking include defense mechanisms (such as absolutism or primary certitude, denial, projection) culturally conditioned assumptions, authoritarianism, egocentrism and ethnocentrism, rationalization, compartmentalization, stereotyping and prejudice.[1]

Many of these critical thinking skills and dispositions coincide with ideas that have been generated since the 1930s by the International Society for General Semantics, its Institute for Propaganda Analysis, and its journal, *Etc*. A general semantics approach to rhetorical invention is implicit in Dumke's reference to "an understanding of the formal and informal fallacies of language and thought"—suggesting questions of denotation and connotation—as well as in the skills in the California guidelines such as distinguishing fact from opinion and identifying stereotypes, "partialities and prejudices in written and graphic materials," bias, propaganda, semantic slanting, different value orientations and ideologies, and the dispositions of facility in perceiving ambiguity and multiplicity of meanings or points of view. The references to value orientations, ideologies, and multiple viewpoints can further serve toward applying a cultural studies perspective examining the ideological implications and subject positions in a wide range of cultural practices and texts.[2]

Teaching the Political Conflicts

The above criteria of critical thinking were incorporated in my article "Teaching the Political Conflicts: A Rhetorical Schema," which was published in *CCC* and later in Tate, Corbett, and Myers's *The Writing Teacher's Handbook* alongside, and as a counterstatement to, Hairston's "Diversity, Ideology, and Teaching Writing." The article outlined units of study in a sophomore-level argumentative and research writing course aimed at understanding (1) the semantic ambiguity, complexity, and subjectivity in public usage of political terms like left-wing, right-wing, liberal, conservative, radical, moderate, freedom, capitalism, socialism, Marxism, and fascism; (2) broad and specific ideological differences between the political left and right, along with the wide range of positions within both; (3) the psychological factors that frequently bias political sources and ourselves as readers and writers, teachers and students, such as ethnocentrism, authoritarianism, rationalization, compartmentalization, defense mechanisms, stereotyping and prejudice (as well as the problematics of what does and does not constitute prejudice in matters of race, gender, and class, including the semantics of terms like "reverse discrimination"); (4) sources and modes of propaganda and politically partisan, biased, or deceptive rhetoric in sources and cultural texts, particularly in the form of special interests, conflicts of interest, and special pleading; (5) means of locating and evaluating conflicting sources on political issues and of incorporating comparative critical analysis of them in papers.

This pedagogical framework necessitates self-critically problematizing one's own subject position as a teacher, researcher, or writer in relation to the positions of one's students and of sources they (and we) research and write about. In this way, the teacher/scholar is free to present her own perspective—not privileged or forced on students as uncontested, monologic truth—but simply as one subjective viewpoint to be understood by students and tested dialogically against opposing ones. The teacher's responsibility in grading students then becomes not whether they have agreed with her views, but simply how clear an understanding of the opposing views and arguments at issue they have expressed in discussion and writing.

In my particular coursework and scholarship, the introduction of socialism as a term in political semantics, defined in opposition to capitalism, enables me to present sources and arguments in support of my own partiality to socialism, while making it clear that this is precisely a partisan viewpoint and assigning students to find the best conservative sources they can to refute these arguments. I also invited UCLA writing instructor Jeff Smith to write a conservative alternative version of "Teaching the Political Conflicts," pointing out the biases he found in my attempt to arrive at an unbiased definition of points of opposition between the left and right. Reading this exchange illustrates for students the necessity for an open-ended, dialectical approach to defining political oppositions, and it provides a prompt for debate among students along similar lines. Students further apply this multiperspectival approach through class debates and research locating diverse sources on the same topic, and in papers summarizing, analyzing, and evaluating the sources' opposing lines of arguments—including a dimension of addressing their own subjective viewpoints on the opposing lines.[3] The four heuristics at the end of this article serve to generate questions for analysis of sources (and checklists for students' own rhetoric) as well as cues for organizing these inquiries into papers.

To reiterate, I have not presented either these heuristics or the broader approach to rhetoric here as definitive solutions to the vexing challenge of devising means of analyzing political rhetoric, and of teaching such analysis, that encompass its infinite semantic complexities, or of the intrusion of subjective biases in any attempt to devise an objective method for this kind of analysis. The very project of teachers' gaining an objective perspective on their own subjective beliefs and grading guidelines can be fiendishly difficult—total objectivity on highly controversial issues being virtually impossible to guarantee. I grade students not on their agreement or disagreement with my own views, but on their success in comprehending and evaluating opposing sources and arguments. With the best will in the world, however, it may still be impossible—and unwarranted—to be totally free from bias in our judgments of students' analyses. All we can do is to be forthright with students about the problem, establish clear criteria for judgment, and be open to dialogue over how successfully we have adhered to those criteria. My experience is that students greatly appreciate bringing this touchy subject out in the open and that they respond fairmindedly to what they perceive as fairmindedness in the teacher.

I suggest that the daunting topic of subjectivity in teaching calls for a whole new realm of rhetorical inquiry, as yet undefined and unexplored in any systematic form, beyond a genre of articles describing postmodernist classroom experiments pooling multicultural student and source subject positions in what usually seems like deliberately nonjudgmental, undirected group therapy (see for example, Ellsworth, Fishman and McCarthy). Berlin's late works like *Rhetorics, Poetics, and Cultures* and Ira Shor's work in Freirean critical pedagogy perhaps come closest to the kind of effort I am calling for—although as much I admire Shor, I take issue with the way he attempts to reduce the influence and authority of the teacher virtually to zero.

I do not believe this avenue of inquiry need end in deconstructionist absolute skepticism or postmodernist relativism denigrating the pursuit of objective facts and denying any standards of critical judgment or higher order reasoning. I believe, rather, that we can best aspire toward objectivity and critical standards through constant scrutiny of the psychological and cultural obstacles in others and in ourselves that impede objectivity, and that this is the most positive lesson of postmodernist theory. My schemas and problematics are quite tentative and open-ended, inviting refinements and alternatives—let a hundred heuristics for analysis of political rhetoric bloom. My highest hope is simply that my efforts will help prompt the development of a significant new field of discovery for scholars of rhetoric.

Appendix: Four Heuristics for Analyzing Political Arguments

Rhetoric: A Checklist for Analyzing Your Own and Others' Arguments

I. When you are expressing your views on a subject, ask yourself how extensive your knowledge of it is, what the sources of that knowledge are, and what restrictions there might be in your vantage point. When you are studying a writer on the subject (or when she cites a source on it), try to figure out what her qualifications are on this particular subject. Is the newspaper, magazine, book publisher or research institute he is writing for (or citing) a reputable one? What is its usual slant (see A Semantic Calculator below, #l and #2)?

2. Are you, as reader or writer (or is the author), indulging in rationalization, or wishful thinking—believing something merely because it is what you want to believe? In other words, are you distinguishing what is personally advantageous or disadvantageous for you from what you would objectively consider just or unjust?
3. Are the actions of the author, or those she is supporting, consistent with her professed position, or is she saying one thing while doing another? (This is one form of compartmentalization, the other most common one being internal inconsistencies in the author's arguments.)
4. Are all of the data ("facts") or quotations correct? Any data used misleadingly or quotes taken out of context?
5. Does she make it clear, either by explicit definition or by context, in exactly what sense she is using any controversial or ambiguous words? In other words, is she using vague, unconcretized abstractions, or is she concretizing her abstractions? Any evasive euphemisms (i.e., "clean" words that obscure a "dirty" truth)?
6. Are the generalizations and assertions of opinion—especially those that are disputable or central to the argument—adequately qualified and supported by reasoning, evidence, or examples?
7. Any unjustifiable (i.e., not supported by adequate evidence) emotional appeal through empty "conditioned response" words (or "cleans" and "dirties"), name-calling, straw man, or innuendo?
8. Are the limits of the position defined or are they vulnerable to being pushed to absurd logical consequences (reduction to absurdity)? In other words, does she indicate where to draw the line?
9. Are all of the analogies (saying two situations are similar) and equations (saying two situations are the same) valid?
10. Does she (dishonestly) stack the deck through using a double standard or selective vision? I.e., is she using half-truths, leaving out arguments, or suppressing facts that might contradict her arguments? Are there faults on her side that correspond to the faults she has pointed out in the opposing position? Or does she (honestly) acknowledge the opposition, fairly balancing all the evidence and arguments of one side against those of the other, giving each side's accurate weight and evaluating them in accurate proportion to each other?
11. Any faulty causal analyses? Does she view any actions as causes that may really be effects or reactions? Any post hoc reasoning—that is, when she asserts that something has happened because of something else, might it be true that the second happened irrespective of, or even in spite of, the first? Has she reduced a probable multiplicity of causes to one (reductionism)? When she argues that a course of action

has been unsuccessful because it has been carried too far, might the opposite be true—that it has been unsuccessful because it has not been carried far enough (or vice versa)?

12. Other logical fallacies, especially evading the issue, non sequiturs (conclusions that don't follow logically from the arguments preceding them, or two statements that seem to be related but aren't), either-or thinking or false dilemmas or false dichotomies?
13. Theory vs. practice: Are the theoretical proposals practicable or the abstract principles consistent with empirical (verifiable) facts and probabilities, and based on adequate firsthand witness to the situation in question? Do the author and her allies practice what they preach?

A Semantic Calculator for Bias in Rhetoric

(This derives from various versions of Hugh Rank's neo-Aristotelian "Intensify-Downplay" schema.)

1. What is the author's vantagepoint, in terms of social class, wealth, occupation, gender, ethnic group, political ideology, educational level, age, etc.? Is that vantagepoint apt to color her/his attitudes on the issue under discussion? Does she/he have anything personally to gain from the position she/he is arguing for, any conflicts of interest or other reasons for special pleading?
2. What organized financial, political, ethnic, or other interests are backing the advocated position? What groups stand to profit financially, politically, or otherwise from it? In the Latin phrase, *qui bono*?
3. Once you have determined the author's vantage point and/or the special interests being favored, look for signs of ethnocentrism, rationalization or wishful thinking, sentimentality, one-sidedness, selective vision, or a double standard.
4. Look for the following forms of setting the agenda and stacking the deck reflecting the biases in No. 3:
    a. Playing up:
        (1) arguments favorable to one's own side.
        (2) arguments unfavorable to the other side
        (3) the other side's power, wealth, and misconduct.
    b. Downplaying (or suppressing altogether):
        (1) arguments unfavorable to one's own side
        (2) arguments favorable to the other side
        (3) one's own side's power, wealth, and misconduct.
    c. Applying "clean" words (ones with positive connotations) to one's own side, without support.

Applying "dirty" words (ones with negative connotations) to the other, without support.

d. Assuming that the representatives of one's own side are trustworthy, truthful, and have no selfish motives, while assuming the opposite of the other side.
e. Giving credit to one's own side for positive events. Blaming the other side for negative events.

5. If you don't find strong signs of the above biases, that's a pretty good indication that the argument is a credible one. Every rhetorician tries to control the agenda by persuading the audience that her side's arguments are more worthy of attention and sympathy than the other's, and this is a perfectly legitimate aim, so long as what is played up and down is supported by good reasoning and evidence, the other side's strongest arguments presented fully and fairly prior to demonstrating why they are less valid or worthy of priority—rather than being suppressed, distorted, or discredited solely by semantic slanting.
6. If there is a large amount of one-sided rhetoric and semantic bias, that's a sign that the writer is not a very credible source. However, finding signs of the above biases does not in itself prove that the writer's arguments are fallacious. Don't fall into the ad hominem ("to the man") fallacy—evading the issue by attacking the character of the writer or speaker without refuting the substance of the argument itself. What the writer says may or may not be factual, regardless of the semantic biases. The point is not to let yourself be swayed by words alone, especially when you are inclined to wishful thinking on one side of the subject yourself. When you find these biases in other writers, or in yourself, that is a sign that you need to be extra careful to check the facts out with a variety of other sources and to find out what the arguments are on the other side of the issue.

Predictable Patterns of Political Rhetoric

Leftists will play up:	*Rightists will play up:*
Conservative ethnocentrism, wishful thinking, and sentimentality rationalizing the selfish interests of the middle and upper class and America abroad	Leftist "negative thinking," "sour grapes," anti-Americanism, and sentimentalizing of the lower-classes and the Third World
Right-wing bias in media and education; power of business interests and administrators	Left-wing bias in media and education; power of employees

Rip-offs of taxpayers' money by the rich; luxury and waste in private industry and the military	Rip-offs of taxpayers' money by the poor; luxury and waste by government bureaucrats; selfish interests and inefficiency of labor unions, teachers, students, and others in the nonprofit sector
Conservative rationalization of right-wing extremism and foreign dictatorships allied with U.S. (e.g., El Salvador, South Vietnam); rightists' use of "Communism" as scapegoat for rebellion against right-wing extremism*	Leftist rationalization of Communist dictatorships or guerillas (e.g., Nicaragua, North Vietnam) and wishful thinking in leftists' denial of Communist influence in anti-American rebellions*
US military strengths, selfish interests of the strengths; military and defense industry; right-wing scare tactics about Communists or other adversaries*	Communists' or other adversaries' manipulation of left "doves"; left-wing scare tactics about nuclear war*

* The oppositions concerning Communism and nuclear war have become largely obsolete by now, but they dominated American public discourse from about 1945 to 1990, and students need to understand them in reading sources from that period. Also, Communism may have been replaced as an issue, but the patterns of argument about it have been, and predictably will continue to be, succeeded by similar patterns applied to whatever new groups are viewed as threats.

Ground Rules for Polemicists

An additional heuristic, which was published in my 1997 article "Ground Rules for Polemicists: The Case of Lynne Cheney's Truth," in application to rhetorical analysis of Cheney's book *Telling the Truth,* comes out of the following refinements on the topic of partisanship and bias in the earlier material.

The *American Heritage Dictionary of the English Language* (1969) defines a polemic as "a controversy or argument, especially one that is a refutation of or an attack upon a specified opinion, doctrine, or the like." Polemics involves strongly opinionated, often partisan arguments, and some scholars and teachers of rhetoric place a negative connotation on polemics, as being opposed to objective or scientific writing. My viewpoint, however, is that polemics in public discourse is a vital subject for study, and that we simply need to judge polemical arguments on how well they adhere to the principles of responsible rhetoric.

The negative connotation often attached to polemics derives partly from the false equation of polemics with invective, defined by American Heritage as a form of argument "characterized by abuse and insult," and even with propaganda. Polemics sometimes does take the form of invective or propaganda, but not necessarily. The following rules for fair play in polemics indicate principles that responsible polemicists honor, but that writers of invective and propaganda do not. As readers and writers, we should not only attempt to follow these rules in whatever we write, but use them to evaluate the sources we are reading and writing about.

1. Do unto your own as you do unto others. Apply the same standards to yourself and your allies that you do to your opponents, in all of the following ways.
2. Identify your own ideological viewpoint and how it might bias your arguments. Having done so, show that you approach opponents' actions and writings with an open mind, not with malice aforethought. Concede the other side's valid arguments—preferably toward the beginning of your critique, not tacked on grudgingly at the end or in inconspicuous subordinate clauses. Acknowledge points on which you agree at least partially and might be able to cooperate.
3. Summarize the other side's case fully and fairly, in an account that they would accept, prior to refuting it. Present it through its most reputable spokespeople and strongest formulations (not through the most outlandish statements of its lunatic fringe), using direct quotes and footnoted sources, not your own, undocumented paraphrases. Allow the most generous interpretation of their statements rather than putting the worst light on them; help them make their arguments stronger when possible.
4. When quoting selected phrases from the other side's texts, accurately summarize the context and tone of the longer passages and full texts in which they appear.
5. When you are repeating a secondhand account of events, say so—do not leave the implication that you were there and are certain of its accuracy. Cite your source and take account of its author's possible biases, especially if the author is your ally.
6. In any account that you use to illustrate the opponents' misbehavior, grant that there may be another side to the story and take pains to find out what it is. If opponents claim they have been misrepresented, give them their say and the benefit of the doubt.
7. Be willing to acknowledge misconduct, errors, and fallacious arguments by your own allies, and try scrupulously to establish an accurate proportion and sense of reciprocity between them and those you criticize in your opponents. Do not play up the other side's forms of power while denying or downplaying your own side's.

8. Respond forthrightly to opponents' criticisms of your own or your side's previous arguments, without evading key points. Admit it when they make criticisms you cannot refute.
9. Do not substitute ridicule or name-calling for reasoned argument and substantive evidence.
10. Do not, as writers of invective and propaganda often do, accuse the other side of committing all of these rhetorical abuses, without presenting adequate evidence that they have done so, simply as a smokescreen to obscure valid arguments from invalid ones.

Notes

1. For a list of citations in support of the previous two paragraphs, see Lazere, "Literacy and Mass Media" and *American Media,* pp. 417-20.

2. Lazere, *American Media* is a collection of critical essays applying this perspective.

3. I have synthesized the lines of argument discovered by my students over several terms during which they focused on controversies over the present distribution of wealth in the United States, in "Teaching the Conflicts About Wealth and Poverty."

Works Cited

Berlin, James A. *Rhetorics, Poetics, and Cultures: Refiguring College English Studies.* Urbana: NCTE, 1996.

California State Board of Education. *Curriculum Guidelines, Grades 8–12.* Sacramento: State Board of Education, 1984.

Dumke, Chancellor Glenn. Executive Order 338. Long Beach: California State University, 1980.

Ellsworth, Elizabeth. "Why Doesn't This Feel Empowering? Working Through the Repressive Myths of Critical Pedagogy." *Feminisms and Critical Pedagogy.* Ed. Carmen Luke and Jennifer Gore. New York: Routledge, 1992. 90–119.

Fishman, Stephen M., and Lucille McCarthy. "Community in the Expressivist Classroom." *College English* 57 (1995): 62–81.

Graff, Gerald. *Beyond the Culture Wars: How Teaching the Conflicts Can Revitalize American Education.* New York: Norton, 1992.

Hairston, Maxine. "Diversity, Ideology, and Teaching Writing." *CCC* 43 (1992): 179–93. Rpt. in Tate, Corbett, and Myers: 22–34.

Lazere, Donald, ed. *American Media and Mass Culture: Left Perspectives.* Berkeley: U of California P, 1987.

———. "Ground Rules for Polemicists: The Case of Lynne Cheney's Truths." *College English* 59 (1997): 661–85.

———. "Literacy and Mass Media: The Political Implications." *New Literary History* 18 (1986–87): 237–55.

———. "Teaching the Conflicts About Wealth and Poverty." Karen Fitts and Alan France, eds. *Left Margins: Cultural Studies and Composition Pedagogy.* Albany: State U New York P, 1995. 189–205.

———. "Teaching the Political Conflicts: A Rhetorical Schema." *CCC* 43 (1992): 194–213. Rpt. in Tate, Corbett, Myers: 35–52.

National Council of Teachers of English. "Resolution on Public Language," 1975.

Rank, Hugh. *Persuasion Analysis: A Companion to Composition.* Park Forest: Counter-Propaganda P, 1988.

Rockefeller Foundation Commission on the Humanities. *The Humanities in American Life.* Berkeley: U of California P, 1980.

Shaughnessy, Mina P. *Errors and Expectations.* New York: Oxford UP, 1977.

Shor, Ira. *Critical Teaching and Everyday Life.* Chicago: U of Chicago P, 1987.

———. *When Students Have Power.* Chicago: U of Chicago P, 1996.

Smith, Jeff. "Response to Donald Lazere's 'Teaching the Political Conflicts.'" Unpublished manuscript, 1998.

Tate, Gary, Edward P. J. Corbett, and Nancy Myers, eds. *The Writing Teacher's Sourcebook.* 3rd ed. New York: Oxford U P, 1994.

American Pragmatism and the Public Intellectual

Poetry, Prophecy, and the Process of Invention in Democracy

JAY SATTERFIELD
FREDERICK J. ANTCZAK

We cannot have rights unless we have common attitudes.

George Herbert Mead, On Social Psychology *(1918)*

Inventional theories today must negotiate a course littered with the ruins of collapsed foundations. To many radical academics, the world has become a complex system of competing meta-narratives, all equally baseless. But there is a risk that those who dismantled philosophical foundations traded their political power for this freedom not to believe. We propose to search the ruins for reasons both to believe and to act—for an anti-foundational inventional theory that offers clear strategies for creating new, politically effective ideas. We find it in an unlikely place—the pragmatic tradition.

Pragmatism may seem like an odd place to search for a powerful inventional theory: Randolph Bourne decried the lack of ethical force in pragmatic thought when he lambasted John Dewey for supporting the First World War (53–64). Pragmatism has also been criticized for being shortsighted incrementalism that lacks a concrete vision of the big picture; and C. G. Prado elaborated on the charges of "nihilism and vicious relativism" (12) laid on pragmatic thought. This essay searches for a pragmatism that begins to address such objections; further, it demands that pragmatism, by its own criteria, should "work." In this case, what it means to work is to provide an intellectual and inventional alternative that leads to a public voice. Even that criterion, however, will prove slippery.

A Search for Common Ground

Pragmatism has always been concerned with how truths are constructed. But this concern has led in two seemingly different directions, most visibly distinguished by very different aims assigned to invention. For neo-pragmatists, the pragmatic tradition of interrogating truth making offers a path for identity formation; it facilitates the sort of invention and indefatigable reinvention of the self that has always appealed to Americans. But pragmatism also offers a tradition that constructs an outwardly directed political framework, oriented—if it can generate a sufficiently powerful mode of invention—at creating social change. Both traditions, we will argue, have roots in William James and John Dewey. But this contrast between an inwardly directed form of invention that seeks to remake the self and an outwardly aimed form that seeks effectiveness in the political arena delineates different commitments. Although these commitments may be complementary, they always work in tension with respect to one another. The former approach is led among contemporary pragmatists by Richard Rorty; defined by a commitment to the self, it looks to great individuals and prophets to enlarge and articulate personal possibilities. The latter, led among contemporary pragmatists by Cornel West, is also defined by a commitment to selves, but selves seen together as rooted in, intellectually and practically implicated with, community. Here, to borrow a phrase from James, is a difference that really makes a difference in the character of invention. According to the pragmatic tradition, "truth is one species of good, and not, as is usually supposed, a category distinct from good" (James 76), so any truth becomes manifest in good acts. Therefore a pragmatic theory of invention is inseparable from a system of ethics. For Rorty, the invention of new truths leads to personal self-actualization. For West, invention leads to social change which eases oppression and suffering.

Rorty's Ideal Space

Much of the current interest in pragmatic thought traces to Richard Rorty, who almost single handedly rediscovered and legitimated the pragmatic tradition that had been so thoroughly discounted by contemporary philosophers. But Rorty's form of pragmatism is inherently individualistic and lacks a sufficiently constructive social element. In perhaps his most characteristic rhetorical move, Rorty focuses his attentions inward, in audience-limiting ways: first inward on the academy, then in parallel fashion inward on the self. For Rorty, there is a strict division of the public and private spheres, and philosophy can only address the private. His turn inward in terms of audience reflects his frustration with the posturing of philosophers in the academy who think their theories are significant beyond their own personal space. When Rorty gave up on foundations and denied the epistemological force of philosophy, he also forever condemned philosophy to the private realm. Rorty's ideas offer inventional space only for the individual self (a kind of free associational arena for constant rediscovery and reinvention of the self), but his ideas become ineffectual when turned outward into a public space.

Rorty is not unconcerned with society; he only denies a public role for philosophers. The public/private dichotomy creates a genuine disassociation of Rorty from Dewey, as well as a disassociation of neo-pragmatism from pragmatism. Dewey wanted society to be built upon the scientific method. In Sidney Hook's reading, "[t]he heart of Dewey's social philosophy is the proposal to substitute for the existing modes of social authority the authority of scientific method," with the public philosopher as the guiding force (151). Dewey believed in the pragmatic notion of experimentalism and wished to see a community of individuals acting together creatively to build a better society. Social progress occurred through scientific experimentation and social engineering based on democratic principles. But while Dewey had faith in the scientific method, he was not a utopian. Dewey's scheme assumed an actively involved polity creating and conducting its own experiments toward a better society—but a society always advancing and always in flux.

Rorty, on the other hand, separates the private from the public in a way Dewey could never accept. Rorty critiques Foucault for wanting to instill our autonomy in institutions:

> It is precisely this sort of yearning which I think should, among citizens of a liberal democracy, be reserved for private life. The sort of autonomy which self-creating

> ironists like Nietzsche, Derrida, or Foucault seek is not the sort of thing that could ever be embodied in social institutions. Autonomy is not something which all human beings have within them and which society can release by ceasing to repress them. It is something which certain particular human beings hope to attain by self-creation, and which a few actually do. (*Contingency, Irony, and Solidarity* 65)

Rorty's brand of neo-pragmatism gives up on the scientific social engineering of pragmatists like Dewey. He wants government to stay out of our business and let us get on with our own autonomous self-creation. Rorty's form of inventional theory revolves around personal self-actualization and abandons the public realm, as evident in a quip from one of his critics concerning his ironic distance from society: "[I]s it inconsistent with the tenets of critical rhetoric, for example, to hold liberal bourgeois views and advocate mixing in the 'bazaar' during the day but retiring to the comforts of one's club at night?" ("Voice of Pragmatism" 17). Such a withdrawal into the personal on key issues like autonomy is not an attitude that Dewey, James, or West could maintain.

Of course, William James and Cornel West approach the issue differently. Neither would go in for scientific rationalism, at least not with Dewey's unbridled enthusiasm. All three (West, James, and Dewey) want institutional change, and all three are united in a willingness to participate actively to effect those changes. Further, West simply cannot look at the plight of the downtrodden and the decay of inner cities and believe that liberalism works to the extent Rorty claims. Put another way, West cannot in conscience retreat from the bazaar to the scholar's study, nor can he reject the importance of public institutions.

Still, there is in Rorty a strain of romanticism that illustrates the potential public, pragmatic voice. The "strong poet" and the "utopian revolutionary" are the heroes of his ideal liberal society. Through personal genius they offer an inventional theory—but a theory that must patiently wait for their arrival. Interestingly, neither the strong poet nor the utopian revolutionary is alienated from society in the traditional sense of alienation: "This is the idea that those who are alienated are people who are protesting in the name of humanity against arbitrary and inhuman social restrictions. One can substitute for this the idea that the poet and the revolutionary are protesting in the name of the society itself against those aspects of the society which are unfaithful to its own self-image" (*Contingency, Irony, and Solidarity* 60). While utopian thought is good and has power, as does the voice of the strong poet, poets and revolutionaries are people trying to change society by acting from within. Here is Rorty's space for incremental change, what James called meliorism; but it exists only in the presence of the genius. It depends to a great extent on an ideal society: "one can define the ideally liberal society as one in which this difference [between the revolutionary and the reformer] *is* canceled out" (*Contingency, Irony, and Solidarity* 60, emphasis in original). But the distance from Rorty's Charlottesville study to "the bazaar," the less-than-ideal, everyday liberal society, remains. In that ironic distance lie the distinctions between Rorty's neo-pragmatism and the pragmatism of public intellectuals, between neo-pragmatic and pragmatic theories of invention. The strong poet does not radically change society's form; he or she only offers new spaces for personal self-discovery within society. This creates two inventional anomalies. First, Rorty does not clarify the role of community in his kind of invention, except as audience construed in a relatively passive, aestheticized sense, whose role is to follow blindly the strong poet. Second, to construe rhetorical invention as ultimately personal puts into question Rorty's position, or at least his centrality, in the pragmatic tradition.

A Space for the Public Philosopher

Rorty has become something of a public celebrity. He is one of the few members of the philosophical community that the general public might recognize because he writes periodically for the *New York Times*. He has, however, stepped away from the role of the public philosopher. It is a move entirely consistent with his belief that philosophy has little or nothing to offer the world. He has turned his attention to individual transcendence, and thereby stands in an important pragmatic tradition of prizing the individual.

To find a truly public philosopher in the pragmatic tradition, we must turn to one of Rorty's star pupils, Cornel West. In *The American Evasion of Philosophy*, West credits Rorty with rediscovering and re-enlivening pragmatic thought, but he accuses his mentor of failing to make pragmatism a truly effective force: "Rorty leads philosophy to the complex world of politics and culture, but confines his engagement to transformation in the academy and to apologetics for the modern West" (207). West wants to take Rorty's historicism out of academic debates and place it in a political world. He asserts: "[T]he goal of a sophisticated neo-pragmatism is to think genealogically about specific practices in light of the best available social theories, cultural critiques, and historiographic insights and to act politically to achieve certain moral consequences in light of effective strategies and tactics" (209). Rorty's failure is his inability "to act politically" or "think genealogically." West charges that Rorty's inward turn has left his philosophy in a polemical battle with academic philosophers—"and hence barren" (207). West moves on to rediscover the side of Dewey that Rorty discards: the overtly political and very public Dewey. Not surprisingly, West sounds more than vaguely reminiscent of Dewey when he calls for "the best available social theories" to be marshaled toward political change. While Dewey opted for the scientific method, West turns toward the social sciences as grounds for social experimentation. What we see in *The American Evasion of Philosophy* is West's attempt to create for himself (and to create himself, almost in Rortyian fashion) a genealogy of thought that allows him to take what he learned from Rorty and utilize it for political purposes. He found in Rorty a historically grounded, anti-foundationalist philosophy, but he must look to Rorty's predecessors for a tradition of public engagement. Interestingly, what he discovers is a very American tradition—one that believes in the power of ideas for social and political transformation.

West is one of the best-known academics of the day. He has been profiled in *Newsweek, Time,* the *New Yorker,* the *New York Times Sunday Magazine,* and even *Vibe.* He publishes widely in academic journals, but also writes for the *New York Times;* and his essays have appeared in journals as diverse as *Spin, Dissent, Z,* the *Nation,* and the *New Republic*. In addition, he is a tireless public speaker who will lecture anywhere, from Ivy League lecture halls, to street corners and coffee shops, to the Million Man March on Washington. He wrote the best-seller *Race Matters* and occasionally appears on television to comment on political and social issues. He sees himself as a public intellectual and relishes the role.

William James offers West a rich pragmatic tradition of public engagement. George Cotkin argues in *William James, Public Philosopher* that James became a public philosopher because it fit with his ideas about pluralism, and it gave purpose to his life. Cotkin says that James was much in demand as a speaker, though he primarily addressed academic groups. The notable exception is his participation in Chautauqua meetings. James's books sold well to a diverse audience, and Cotkin cites a study conducted by the St. Louis Public Library from 1931 to1934 that showed James's writings were still "being read by a diverse audience composed of 'a trunk maker, machinist, stenographer, retired farmer, clerk, three wives, two physicians, a salesman and a post office worker'" (12). While James lobbied against a bill that would have required medical licensing for doctors, his main political activities revolved around his anti-imperialist stance. He was a frequent participant in the functions of the New England Anti-Imperialism League, and his famous essay "The Moral Equivalent of War" appeared in several popular magazines (including *McClure's*).

It is almost redundant to use "public" and "Dewey" in the same sentence. Sidney Hook calls Dewey "the philosopher of the plain man" and lists several pages of Dewey's public/political activities (17, 19–23). Dewey helped to form the ACLU, and he was a national leader of education reform. He helped organize the Farmer-Labor Party, helped draft "The Plan to Outlaw War," publicly defended Sacco and Vanzetti, worked to see that Trotsky got a fair asylum hearing, and headed a Commission of Inquiry into the Moscow Trials. He helped organize the Teachers Guild and was an important member of the Teachers Union. He was chairman of the People's Lobby. He is probably most famous for his ill-fated defense of American involvement in World War I.

The Prophecy of the Public Philosopher

The issue of public engagement is crucial for understanding a pragmatic theory of invention. For West, truths get invented in public forums. Because "truth is in the making" (as James put it), what a democratic anti-foundationalist like West or Dewey regards as coming closest to capital-T Truth is what emerges in the negotiating process among at least some of the interested parties. A pragmatic inventional theory rests on the notion that politically effective knowledge must be created in an historically contingent public space. The pragmatic theory of invention emerges from this basic attitude toward the nature of truth and knowledge. West's commitment to invention is visible insofar as he is committed to empowering more of the parties whose interests are at stake to have an active role, a voice, in the negotiation. James, Dewey, and West see themselves as facilitators of the dialogue necessary for invention; this is a different kind of "prophecy" from the monological poetry of the strong poets—a contrast of West's vision of incremental change as over against Rorty's utopian visions. But, for West, the possibility of meaningful action depends on two interrelated things: community and faith. West can offer a somewhat greater measure of the latter because he works with a richer dimension of the former, a theory of invention turned outward toward the community. Pragmatic invention for West is prophecy, not poetry, addressing mundane, prosaic exigencies for particular communities. But prophecy only contains the seeds of truth. Prophecy alone does not constitute invention—it must be nurtured through the dialogue of an engaged public.

What is intriguing, and problematic, is that West personally rests his anti-foundationalism on a deep Christian faith (a traditional favorite of many foundationalists). Christianity and the humanism embodied in the Gospels make pragmatism "work" for West. Rorty needs no such faith, or at least it is not central to his philosophy, and he goes further to reject Enlightenment thought because it rests on a foundation that claims natural human rights. This is not to say that Rorty thinks we have no human rights; it only means that he does not find, or find necessary, any foundation for those rights. Humanism, for Rorty, is just another foundational meta-narrative. Rorty continues to hang on to humanism without a base; in his view, liberal democracy does not need the base because it just plain works. For West, faith is central, but in a sense it remains anti-foundationalist because a plurality of faiths exists (though they may need that common thread of humanism to tie them together). West titled one of his collection of essays *Keeping Faith,* but the faith he refers to is more along the lines of a faith in the possibility of some sort of progress—a faith that is expressed in words and manifested in works. This faith separates the James-Dewey-West pragmatic tradition from Rorty. But is faith in progress and humanism pragmatic, or does that faith form a foundation in an anti-foundationalist philosophy? The reliance of West, James, and Dewey on the humanist impulse could be argued to form a Truth—a truth that only Rorty is willing to give up.

Elements of Pragmatic Invention

Cornel West does not see himself as a rhetorician, and he never has felt a need to clearly explicate a theory of invention. We can extract a theory of invention from his philosophy as stated in *Beyond Eurocentrism and Multiculturalism.* Through a juxtaposition of that theory with West's commentaries on the Million Man March, we can piece together a theory of invention and then examine how it becomes manifest pragmatically.

West's concept of invention depends on the "prerequisites" of prophetic thought for inspiration, on the "elements" of pragmatic thought for guidance and organizational superstructure. These refigure invention: for West, invention focuses less narrowly on a moment of creation of arguments, but rather implicates a whole persuasive process that opens venues for new expression. He adds a deeply communal element to develop a more "dynamic means for the cultivation and encouragement of the potentialities and possibilities of unique individuals" (*Beyond Eurocentrism* 168).

In *Beyond Eurocentrism and Multiculturalism,* West lists four prerequisites for prophetic thought: *discernment, human connection, tracking hypocrisy,* and *hope. Discernment* is "the capacity to provide a broad and deep analytical grasp of the present" (3). It reaches beyond theory to involve critical capacities. West insists on a strong sense of history and a grasp of the complexity of human situations. *Human connection* folds into invention the power of empathy, described by West in distinctly rhetorical terms. West describes empathy as "the capacity to get in contact with the anxieties and frustrations of others." West mixes elements of his Christianity and universalism with a dash of old-fashioned humanism. "Always attempting to remain in contact with the humanity of others," his prophetic thought eschews blame and resolutely keeps open the possibilities of cooperative action. *Tracking hypocrisy* means "accenting boldly, and definitely, the gap between principles and practice, between promise and performance, and," he adds, "between rhetoric and reality." We need words, from the strong poets and from the rest of us. But rhetoric by itself is not enough; it may even be a deception. Prophecy must comprise both words and deeds, must invent both eloquence and ethics. West emphasizes the ethical aspect by characterizing this prerequisite as always a *self*-critical activity because "we are always complicit with the very thing we are criticizing" (5). Finally, *hope* means that we must believe that what we do matters.

To address postmodern circumstances, West plays these prerequisites for prophetic thought against three basic elements of pragmatic thought: voluntarism, fallibilism and experimentalism. Voluntarism is selves willingly acting. Although West often wields this concept of unique selves—one is tempted to say "souls"—in a way that hardly sounds postmodern, these selves are also an accretion of experience and thought. Fallibilism is the recognition that the future is risk-ridden and open to revision. Experimentalism revolves around a willingness to act without undue fear either of failure or of the new. In this he follows James, who held that all experience is experiment. The common ground of pragmatism is that "unique selves acting in and through participatory communities give ethical significance to an open, risk-ridden future" (43). Prophetic thought is what happens on that common ground.

West claims that prophetic thought "depends on our capacity to preserve, cultivate, and expand traditions of critique and resistance" (59). These are capacities of discernment, of seeing complexity and drawing on history for inspiration and understanding. Lincoln, Whitman, Emerson, and Du Bois offer traditions of cultural criticism and resistance that we can draw on today. But what is it to cultivate these traditions? First and foremost, it means a commitment to speaking up (which, for West, also means acting up), for making spaces for prophetic thought and making them ring with dialogue. West praises rap and hip hop for opening new venues for expression. He finds Spike Lee's movies flawed, but recognizes how they serve some of the same functions as Baldwin's essays once did for spurring debate. Regardless of the aesthetic quality of *Do the Right Thing,* it opened a space for dialogue about critical issues that needed to be talked through. Such public discussion makes for a more conscious America; and a more conscious America is, for West, an America more prone to action.

Prophetic thought, however, leads West in at least one direction that seems to ignore some of the historicism he so dearly requires. The action he hopes for is not simply the magnificent gesture of the great leader. In responding to problems in the black community, he wants Americans to build more organizational structures for resistance. Effective leaders, he claims, build organizations bigger than themselves. Leadership requires a subjugation of personality, which is as unlikely, West thinks, as Jesse Jackson being eclipsed by the Rainbow Coalition: "His brilliance, energy, and charisma sustain his public visibility—but at the expense of programmatic follow-through. We are approaching the moment in which this style exhausts its progressive potential" (*Race Matters* 67). But West, following James, ties his notion of leadership to the reinvention of the Progressive habits of coalition and alliance. Baseball serves as an example: West notes how well different races get along on the baseball diamond because they have joined together as a coalition in the context of struggle. Ideal products of prophetic thought (like dialogue, organizations, and alliances) can generate a social movement where ideas are tested and actualized. Only then, in a pragmatic perspective, do they become meaningful and realize their truth.

It is nothing new to claim that invention emerges from dialogue. The United States has always been a nation committed to public dialogue.

What West adds here is a kind of superstructure around that dialogue. The superstructure begins in the institution of democracy, includes the organizations of people bent on actions, then reinforces itself with the alliances that allow for inter-group cooperation. So dialogue creates meaningful truths, but it does so within politically minded institutions.

For West, invention comes out of dialogue and collective action. We create new truths, invent new ways of living through a consensual, dialogic approach to action, and democracy is the key institution to foster this process. So far, West stands firmly in the Dewey-James pragmatic tradition. But his most original contribution to pragmatic thought is also a part of the inventional process. West wants to infuse pragmatic thought with a sense of the tragic. Lincoln is his key example: when Lincoln refused to demonize the Southern people in his conciliatory Second Inaugural Address, he kept open the possibility for dialogue. According to West, if each individual recognizes the tragic, sees that evil is in the world as a result of tragedy, and sees himself or herself implicated in the tragedy, there is always hope for a meaningful dialogue that can change things.

In October 1995, West sought just such a meaningful dialogue by participating in the controversial Million Man March, an action he felt he must defend in a *New York Times* editorial. There he claims that he is working from the tradition of Martin Luther King Jr. by opening dialogue with people he has strong disagreements with in order "to promote black operational unity." West is quick to disassociate himself from Louis Farrakhan, who embodies "black hatred and contempt for whites, Jews, women, gay men, and lesbians." But he feels that the media has wrongly portrayed the march as Farrakhan's march and not as a show of black unity. So his second reason for participating is partly an effort to shift media attention from Farrakhan toward black suffering—which he sees as the basis of the march. West's third reason is crucial, he says: "I must march because the next major battle in the struggle for black freedom involves moral and political channeling of the overwhelming black rage and despair." He goes on to say "[y]oung blacks are hungry for vision, analysis and action: radical democrats must go to them and be with them." West sees himself as the carrier of a vision to channel black rage. Farrakhan, the organizer of the march, may be a xenophobic, anti-Semitic, homophobic zealot, but West feels his own presence may meliorate, even channel the emotions of the march. He concludes by saying: "I believe that if white supremacy can be reduced to a minimum, then patriarchy, homophobia and anti-Semitism can be lessened in black America. If I am wrong,

America has no desirable future. If I am right, black operational unity need not preclude multiracial democratic movements that target all forms of racism and corporate power. Whether right or wrong, I must fight. So I march" ("Why I'm Marching"). The organization is bigger than the man. West sees the Million Man March as a moment of black (male) coalition building—not as "Farrakhan's March."

We see many of the elements of West's pragmatic mode of invention played out in his discussion of the Million Man March. First we see his admission of fallibilism—he does not know if he is right or wrong, but he is willing to act and then judge the outcome on a pragmatic level based on results. We also see his desire to create a superstructure to channel black rage into constructive action. He is moving toward creating a coalition to establish black unity. Most impressively, we see an acknowledgment of his sense of the tragic. He refuses to demonize Farrakhan, who stands for many of the evils of society that West is firmly committed to alleviating. By not demonizing Farrakhan, he allows communication to occur and opens up the possibility for a meliorist effect. The march is like a Spike Lee Joint: it may be flawed, but it gets people talking. West refuses to reject the movement because of its leader.

But did it work? It appears not. The march created a flurry of short-term activities but may have failed to produce long-term coalitions. Surely a strong voice of toleration bent on action was present and participating in open dialogue with other concerned citizens (both at the march and through the pages of the *New York Times*). One of West's most positive qualities is his praiseworthy willingness to define exactly where he stands on issues, then openly discuss the issues with whomever he can. He clearly states that he is "a radical democrat devoted to a downward redistribution of wealth and a Christian freedom-fighter in the King legacy—which condemns any xenophobia, including patriarchy, homophobia and anti-Semitism" ("Why I'm Marching"); this is a stance frankly opposed to Farrakhan. But he shows a willingness to engage in dialogue to effect change through the invention of new truths. West felt he needed to represent the voice of the Christian Prophet at the march. Ignoring the march was ignoring a possibility to make a difference.

But does West's theory underestimate the centrality of the charismatic leader in American history? He says he wants institutions to become the focal point for change, and his theory of invention rests on that occurring—but is it possible? West claims that the white media made the march Farrakhan's march, and he hoped to reclaim a part of it through his participation. It seems unlikely that by participating he actually laid claim to any part of the march so much as Farrakhan managed to co-opt West's charisma. If it is only the charismatic leader who must effect, or at least ignite, change by building an institution centered on the leader, then West is dead wrong, and it is Rorty's strong poet or utopian revolutionary (in this case in the form of Louis Farrakhan) who wields the true power of social and institutional invention.

A Pragmatic Test

Invention, in a pragmatic definition, is the creation of new thought that is workable, but also sharable. In other words, pragmatism pushes new thought beyond a speaker's charisma. It seems only fair, then, to ask of West's theory of invention: "Is this pragmatically sound?"—that is, does it work? Does it create public meanings and community actions that change things for the better? By these criteria, the theory seems better developed than its plan of implementation. What West lacks is a clear notion of how the long chain of democracy, dialogue, and sustained collective action can come about. It is clear from his own activity that West has a deep faith in the capacity of oral communication to motivate people; but he never demonstrates that it truly can make a difference. How do we get people organized and form alliances? Aren't there conflicts between such alliances; don't their existences sometimes lead to new conflicts? What of the tension between developing unique individuals, collective institutions, and actions? And last, how can he talk to those who do not share his sense of the tragic and will not listen?

His theories seem well adapted to an earlier historical period, say the America of the 1930s through the 1960s when the left actively formed coalitions. Coalition building is not a habit among the contemporary American left (and it is the left that is most devoid of foundations). The one sphere where coalition building has been most effective in the past decade is on the right, most specifically the religious right where claims of capital T-Truths and capital F-Foundations abound. West would find many of the right's political views and actions deplorable, but he would have to applaud its means. The Christian Coalition has successfully constructed organizations that are separate from the personalities of the leaders. Without foundations, how likely are organizations divorced from charismatic leaders? Is it the strong poet who becomes the foundation, if only on a historical and contingent basis?

The political left has been more fragmented by the original problems of this paper, the collapse of foundations and Eurocentric thought. So the question is, does West provide anti-foundationalists an inventional theory that can recreate a common world that admits coalition building? If not, the whole pragmatic enterprise falls apart. Perhaps what West's inventional theory lacks is an adequate audience analysis for those without foundations. He needs a clear articulation of how to bring people together —where Rorty offers the strong poet, West can only offer faith. His theory of invention needs a culturally effective way to instill attitudes of pragmatic and prophetic thought in people; otherwise, he preaches only to the believers.

Further, is this a problem with West or a problem with pragmatism (and not with the neo-pragmatism outlined by Rorty)? For all the shortcomings of Rorty, he understands the power of charisma. West should learn from Rorty by way of Farrakhan: powerful individual voice is an essential ingredient for sustaining and supporting organizational superstructures, especially with the American audience West does most of his work trying to change. Charismatic leaders provide historically contingent foundations, and only the most effective leader routinizes his or her charisma into the very structure of the organization. Is it pragmatic in addressing a post-foundational audience to downplay individual leadership? West charges Rorty with a kind of nonchalance for backing away from the reality of social problems, but persisting in naive hopes, undisciplined by audience analysis, may be another form of nonchalance.

Works Cited

Bourne, Randolph. "Twilight of Idols." *War and the Intellectuals: Essays by Randolph S. Bourne. 1915–1919.* Ed. Carl Resek. New York: Harper, 1964.

Cotkin, George. *William James, Public Philosopher.* Baltimore: Johns Hopkins UP, 1990.

Hook, Sidney. *John Dewey: An Intellectual Portrait.* New York: Day, 1939.

James, William. *Pragmatism: A New Name for Some Old Ways of Thinking.* New York: Longmans, Green, 1931.

Langsdorf, Leonora, and Andrew R. Smith. "The Voice of Pragmatism." *Recovering Pragmatism's Voice: The Classical Tradition, Rorty, and the Philosophy of Communication.* Albany: State U of New York P, 1995. 1–19.

Mead, George Herbert. *On Social Psychology: Selected Papers.* Ed. Anselm Strauss. Chicago: U of Chicago P, 1977.

Prado, C. G. *The Limits of Pragmatism.* Atlantic Highlands: Humanities P, 1987.

Rorty, Richard. *Contingency, Irony, and Solidarity.* Cambridge: Cambridge UP, 1989.

West, Cornel. *The American Evasion of Philosophy: A Genealogy of Pragmatism.* Madison: U of Wisconsin P, 1989.

———. *Beyond Eurocentrism and Multiculturalism.* Monroe: Common Courage, 1993. Vol. 2 of *Prophetic Thought for Postmodern Times.* 2. vols. 1993.

———. *Race Matters.* New York: Vintage, 1994.

———. "Why I'm Marching on Washington." *New York Times* 14 Oct. 1995, sec. 1: 19.

Inventing Chinese Rhetorical Culture

Zhuang Zi's Teaching

Haixia Wang

Each year, especially around its anniversary, the memory of the tragedy at Tiananmen Square in 1989 is rekindled. To better understand this event, I turn to studies in rhetoric and composition, especially rhetorical invention. If language is indeed power, what kind of power was exercised through the public discursive exchange during this event? What can we learn, through the power of language, about the event and about the public speakers? How was this power implicated in relation to the loss of innocent lives in Beijing? And how can discursive practices help avoid re-occurrences of such a tragedy? These are important questions for the Chinese democratic movement, China studies in general, and comparative/contrastive rhetoric. One issue central to all these questions, and for all the interested groups, is whether democracy is possible in China, a country where democracy has never flourished in its long history. One main argument of this paper is that despite the lack of a democratic political system, some long-established cultural traditions do foster a democratic spirit and thus the potential for helping to invent a modern-day rhetorical culture in China.

In this essay, I use certain key Western rhetorical concepts and the rhetorical thoughts of ancient Chinese Taoist Zhuang Zi.[1] Even though, like democracy, the discipline of rhetorical studies in the Western sense did not exist in ancient China, I find Zhuang Zi's concepts radically democratic and rhetorical. For over two millennia, marginalized Taoism has contributed to the shaping of the Chinese culture as much as mainstream

Confucianism, and the traces of its influence are as ubiquitous. Together, Zhuang Zi's rhetorical thoughts and Taoism's lasting and penetrating influence give Zhuang Zi the potential to help us invent Chinese rhetorical culture today.

Does Zhuang Zi Make Sense?

Zhuang Zi has been interpreted as a relativist, an irrational person, an anti-rhetorical philosopher, as well as a rhetorician. Recent studies both by Chinese and by Western scholars, however, have constructed better understandings of this seemingly unconventional thinker. In 1962, for instance, Robert Oliver made a good case that Zhuang Zi had important rhetorical implications, and in 1989 W. A. Callahan argued that Zhuang Zi's concept of spontaneity [*ziran*], etymologically, meant *discourse,* thus pointing to *ziran*/spontaneity's democratic political implications. The late linguist and sinologist A. C. Graham has also left us with numerous valuable translations and interpretations of Zhuang Zi. In fact, one of the most encouraging and fruitful areas of Zhuang Zi scholarship today is the study of his idea of language use and of rhetoric. Drawing upon these studies, this essay argues that Zhuang Zi, when read rhetorically, makes profound sense for inventing Chinese rhetorical culture for today.

The word *rhetorically* requires, when we face the challenge of ancient Chinese thinkers like Zhuang Zi, not only our knowledge of historical facts but also our understanding of ancient Chinese philosophical thoughts in order for us to draw intelligible and useful lessons from those sophisticated debates of the day. We cannot isolate words and sentences. Interpreting texts rhetorically also requires painstaking attention to nuances—never overlooking the possible significance of minute differences. The following are some examples of how Zhuang Zi can be baffling at first sight, but also how Zhuang Zi's words can be interpreted rhetorically.

Zhuang Zi once favorably quoted the Chinese sophists as follows:

> "None under heaven is more huge than the tip of autumn feather." and "The huge mountain is called small."
>
> "None is more long-lived than the child died-young." and "Our Forefather Peng is called he who has died too young."
>
> "Heaven, earth, with I myself are born at the same time," and "myriads of things with I myself make one." (trans. Wu 161–63)

An isolated reading of these words could lead to the conclusion that Zhuang Zi is a total moral relativist or that these words are unreasonable or irrational. Further, in Zhuang Zi, this view is repeatedly enforced by vivid stories, one of which is the well-known "Morning Three." Laughing at people who labor over making differences the same without knowing that they are at once and inseparably the same and different, Zhuang Zi says: "What is called 'morning, three?' Mr. Monkey-keeper, giving-out chestnuts, said, 'Morning, three and evening, four, all right?' The multitudes of monkeys were all angry. He said, 'If so, then morning, four and evening, three.' The multitudes of monkeys were all pleased" (trans. Wu 142). Is Zhuang Zi a total relativist? Reading Zhuang Zi rhetorically, mathematical logician Raymond M. Smullyan, in his *The Tao Is Silent,* says that Zhuang Zi is simply explaining the nature of the Tao. By implication, "finding the way" is predicated upon the assumption that there is "the way" and there is "the non-way"; the way is not the same as just any way; and one thing, therefore, cannot be just as good as the next. Smullyan painstakingly and successfully distinguishes obeying the Tao from being in harmony with the Tao (37), and this further demonstrates the right and the not-so-right ways of pursuing the Tao. Obeying the Tao, Smullyan explains, one may laboriously, if not fearfully, make efforts to obey rigid distinctions, often made by others, as if the differences were permanent and ultimate; but, Smullyan says, this is not Zhuang Zi's understanding of the Tao. According to Smullyan, Zhuang Zi's understanding is that being in harmony with the Tao means that one spontaneously responds to life moment by moment and therefore sees differences as fluid and provisional.

Smullyan's interpretation opens up the dynamic/rhetorical dimension of Zhuang Zi's vision of the world, a world that is constantly in motion with endless possibilities. Using Smullyan's insight, we can see Zhuang Zi as a thinker who views human life and thus rhetorical invention as analogically, intimately, and also dynamically related to its surroundings. Ordinary perceptions of big and small, long and short, differences and similarities are used by Zhuang Zi to make the point that in different life moments, something huge, long, or similar can become small, short, and different and that differences have a protean and contingent nature. Here, another rhetorical characteristic of Zhuang Zi's thought is that the Tao, the right way, cannot be carbon copied from one life situation to another and therefore the "right ways" are impossible for some wise persons to prescribe for the general public. The Tao of each life situation, with all its potential, can only be discovered by those who are actually involved when they are actively engaged in it.

A. C. Graham has made similar efforts to refute the relativist reading of Zhuang Zi. His studies emphasize how without being a relativist Zhuang Zi manages to collapse the binaries of similarities and differences, unity and diversity. Like Smullyan, Graham takes pains in making important distinctions, such as the distinction between irrationalism and anti-rationalism. In several of his studies, Graham makes it clear that irrationalism is unreasonableness, and anti-rationalism is reasonable protest against a relentless fixation on rationality, in spite of the objective world, common sense, and common beliefs—even in spite of oneself. Zhuang Zi, as Graham points out in *Disputers of the Tao,* is not irrational but rather anti-rational, neither totally relativist, irrational, nor anti-systematic: he himself relies on analytical reasoning as his tool (176).

The next natural question for Graham to explore carefully is to what extent Zhuang Zi relies on reasoning, systematizing, or sorting. Zhuang Zi says: "What is outside the cosmos the sage locates as there but does not sort out. What is within the cosmos the sage sorts out and does not assess. The records of the former kings in the successive reigns in the Annals the sage assesses, but does not argue over" (trans. Graham 57). In the name of Heaven and the Tao, for instance, Zhuang Zi's opponents, such as Confucians, acknowledge diversity within the Tao, thus the necessity of analyzing differences within the unity; Zhuang Zi has no argument with the necessity for analysis. However, Confucians, to Zhuang Zi, carry their power of systematizing to the extent that they attempt to articulate rules and formulate rituals in an effort to perpetuate certain differences, orders, and systems; and this Zhuang Zi sees as futile meddling with the dynamic nature of things. In contrast, therefore, Zhuang Zi's sorting and assessing produce only provisional and contingent categories and evaluations. Harmony with the Tao is beyond the predications the best human reason alone can master; at the same time, human beings are not hopeless victims of fate. Graham notes that Zhuang Zi "never does say that everything is one, always puts the thought subjectively, as the sage treating as one" (*Disputers* 181). In his book *Zhuang Zi: The Butterfly as Companion,* Kuang-Ming Wu concurs with Graham on this issue. This observation confirms that Zhuang Zi does not negate either the need for sorting and classifying or the apparent differences in life—differences between big and small, right and wrong. Again, Zhuang Zi does not view human beings as helpless victims of chance. For him, sages, people in harmony with the Tao, are whole human persons whose powerful reason always functions in balance with other human senses such as intuition and emotion.

We see now why Zhuang Zi should be read rhetorically. A contextualized, careful, and knowledgeable reading helps us understand the rhetorical nature of Zhuang Zi and often enables us to achieve different interpretations from the conventional ones. For instance, contrary to the interpretation that Zhuang Zi claims everything is the same, a remark that can hardly be taken seriously, the result of a more rhetorical reading actually affirms the opposite: Zhuang Zi carefully studies and articulates the complexity of the nature of differences because differences are an important subject of study. Neither his warning against the danger of a simplistic perception of differences nor his emphasis on the importance of understanding the forever shifting boundaries of those differences negates the existence of differences. Similarly, it is the fixation of categories, not the act of sorting, analyzing, evaluating, that Zhuang Zi speaks against.

Even though it is true that Zhuang Zi "preferred private life to office" (Graham, *Disputers* 170) and he was never interested in writing about how to govern, his words imply rhetorical principles that have social-political implications. As we shall see in the following discourse analysis, Zhuang Zi's insight into difference and unity offers an important lesson for us to learn from Tiananmen Square in 1989—a lesson that is anything but irrational or morally relative.

A Contemporary Example of Zhuang Zi's Rhetorical Theory

The antagonism between the two major forces, the army and the students, at Tiananmen Square in 1989 is indeed beyond question; yet, the differences are far from clear-cut, straightforward, and least of all, absolute.

The confrontation took place on June 4, 1989. Between the fourth and ninth of June, the *People's Daily,* the Communist Party's newspaper, was silent about the event, forcing the readers of the newspaper to wait for a decision to be handed down to them. On June 10 the Party leader's speech was printed. The speech contained the following messages: (1) condolences for the lost soldiers' lives; (2) explanations that the event actually took place at a good time (since the veteran war generals were able to command the military forces); (3) official labeling of the event as counterrevolutionary; (4) an excuse for the delayed action of the military (the number of people involved, domestic and abroad, was unprecedented); (5) praise for the military; (6) a proclamation of the importance of political education in the future; and (7) praise for the noble nature of quelling the reactionaries. Research shows that from June 10, the day this speech appeared in the *People's Daily,* until August 31, every entry in the paper's editorial and political discussion sections was a discussion of one or two of these seven themes. Therefore, the meaning of this event was constructed by a benevolent sage for the general public who had no choice but first to wait, then to listen, and finally to study this meaning to reach the sage's level of thought.

This official discourse deployed a powerful audience strategy. The message explained that the reason the government did not act right away was that resistance was unprecedented. The leader maintained that the goal of the students was to "topple our country . . . and to establish the Western dependent bourgeois republic" (trans. Oksenberg 378–79). Further, the leader mentioned the contradiction between the U.S. government's attitude toward Tiananmen in 1989 and the U.S. government's actions at Kent State University in the 1970s. Obviously, the last reference was intended to embarrass the U.S. government but also to appeal to the Chinese audience. It encouraged the Chinese to agree with the government by discouraging them from agreeing with the democratic movement and especially its suspicious foreign connections. The Chinese century-old wariness of the West was invoked; since the West was still very much an unknown to most Chinese, the Chinese audience was encouraged to be cautious about things Western and foreign.

On the other side, the students' discursive practices bore important similarities to the official one: these rhetors seemed to be experienced in manipulating the audience, and their language decided for the audience what they were to think. In their Declaration of Hunger Strike, the students prided themselves on their "purest feelings of patriotism": "We endure hunger, [and] we pursue truth. . . . Democracy is the highest aspiration of human existence; freedom is the innate right of all human beings. But they require that we exchange our young lives for them" (trans. Han 201). This was a strategic move on the part of the students to establish their credibility with the people, especially the government. Two other assumptions were shared by the students and the government: an undemocratic conception of the role of the elite in China and an arhetorical perception of the role of the populace in China's social-political life. Chinese college students have always had a rather complex relationship with the power structure.

The imperial examinations for students in China continued one of the world's oldest and most complete systems of meritocracy, a system which trained and selected civil officials for the imperial court through contests of formulated and formulaic writings. This process of creating a special class has a long history. In his *Chinese Democracy,* political scientist Andrew Nathan, when telling the story of a well-known fourth-century poet, Qu Yuan, comments on the intellectual class's allegiance to the central power structure—an allegiance that is so quintessential that many Chinese take it for granted. His words could just as well describe the students' discourse:

> Tradition of remonstrance in China is ancient. . . . Tradition insists that the remonstrators were always unselfish. They never spoke to protect their personal rights; on the contrary, they put life and property at risk to awaken the ruler to his own interests and those of the state. When remonstrance failed, the act of self-sacrifice affirmed the moral character of the state and set an example for later generations of the minister's duty to guide the sovereign. "He who restrains his prince," wrote Mencius, "loves his prince." (24–25)

Nathan maintains that traditionally Chinese literati were indeed regarded and treated as part of the power structure, not part of the ordinary people. He further contends that the students, holding the same view of government as the Party, did not seem to be aware that such a social change would be incompatible with their traditional relation to the government. Several China specialists agree with Nathan that the traditional Chinese intellectuals' loyalty to the prince was clearly everywhere at Tiananmen. For instance, Elizabeth Perry observes that the students staged public negotiations, even though they were ignored like little children:

> Three student representatives [were] kneeling on the steps of the Great Hall of the People, the one in the middle holding a scroll of white paper high above his head, hoping in this way to move the "servants of the people" into emerging from the great Hall to talk with the students. One minute after another passed. The student representatives became tired—so tired that they sagged down to the ground. Yet they continued to hold the petition above their heads. Approximately fifteen minutes passed; still they were ignored. (63)

The students also insisted that the Chinese working class be kept out of this political and historical event, or at best remain on the margins. They used student IDs as passes to the square; they turned down the workers' repeated offer to go on strike in support of the movement. According to another China specialist, moral philosopher Henry Rosemont, most of the students were gone by the time the tanks rumbled into the square.

Rosemont also points out in his book *A Chinese Mirror* that no student executions were reported by 1991, the year this book was published, although at least forty-two workers' executions had been reported (22). If indeed, ordinary citizens were used as scapegoats by the government, did the students treat them differently? The students' statue of Goddess of Democracy and their Declaration of Hunger Strike were indeed heroic acts, but they were so foreign to ordinary Chinese and so resonant of the Statue of Liberty and the Declaration of Independence that it is obvious that the students considered their audience to be the Chinese and American *governments,* making the ordinary Chinese citizen irrelevant and dispensable.

As Zhuang Zi says, many times differences that appear absolute can actually also be seen as similarities. There were at least two levels in the relationship between the students and the government in Tiananmen. At one level, the students were indeed heroic in their passionate and selfless pursuit for freedom and for democracy, which has forever changed China and will always be admired and remembered. In this respect, they undoubtedly differed from the government. The other level, however, reveals an opposite relationship. At this level, the students are seen as undemocratic, as elitist, in their discursive practices and as determined as the government to keep ordinary people out of social and political decision making processes. In this respect, they are not seen to be fighting for a democracy that was to be lived and participated in by the Chinese people; instead, their goal seemed to be some esoteric theory or some kind of intellectual privilege that was beyond and could be corrupted by the involvement of ordinary Chinese people. This mentality is consistent with centuries-old dynastic discursive formations. Unaware of the significant similarities within the apparent differences between them and the government, students forced the factory workers as well as many others out of the square and out of the effort for democracy. These interpretations are all theorized by Zhuang Zi's notions that there can be essential similarities among differences and that diversity has a complex and dynamic dimension.

The Tao, Spontaneity, Myriad Things, and Rhetoric

As with other ancient Chinese schools of thoughts, the Tao is central to Zhuang Zi's thought and to his caution against the harmful consequences of viewing temporary differences as permanent. Modern China specialists differ in their interpretations of the Tao, as did the ancients; yet some recent studies do agree that the Tao is more like the Western notion of probable truths than of the absolute truth (Hansen, Ames, Hall, Graham, Rosemont, for example). In fact, as many have pointed out, Chinese culture in general has the tendency to focus on a divisible whole, unlike some Western traditions, notably that of Plato, that start with the individual and then move on to aggregates of individuals. In his book *Language and Logic in Ancient China,* Chad Hansen points out: "The mind is not regarded as an internal picturing mechanism which represents the individual objects in the world, but as a faculty that discriminates the boundaries of the substances or stuffs referred to by names. This "cutting up things" view contrasts strongly with the traditional Platonic philosophical picture of objects which are understood as individuals or particulars which instantiate or "have" properties (universals)" (30). That the Tao refers to probability means that rhetorical and democratic ideas existed in ancient Chinese thoughts. In rhetorical terms, some ancient Chinese scholars understood the notion of probable/contingent truth and of the situational contexts (kairotic moments), two important concepts in Western rhetoric.

Related to all this is Zhuang Zi's emphasis on the dynamic relation between the whole and its parts. The Chinese tendency to emphasize the divisible whole, as discussed so far, has been compared to division or whole/parts thinking in the Western rhetorical and topical tradition. Zhuang Zi's concept of the sage helps us to understand that the Tao is not the same as the absolute Truth but the Way or Path. Sages are any human beings who have achieved wholeness as human persons. They use their reason but value their other faculties and abilities just as highly. Similarly, this completeness of human beings, of human environment, and of the entire world is the Tao of all. Myriads of things are indispensable parts in this Whole, the Tao. These parts are not in the whole to fulfill some teleology; their existence, actions, and efforts shape the Tao as much as the Tao shapes them. An individual person's relation to other things and beings, consequently, is as important as his/her relationship to the Tao. Further, even though they are parts of the whole, they do not possess exact characteristics, but are constantly being made by all the forces in the whole.

One consequence for Tiananmen is that only the people involved could discover the Tao, the Path, the Way of democracy during their own moments of Chinese democracy. As valuable as other models and experiences of democracy in the world are, Chinese democracy can only be constructed on Chinese indigenous practices and beliefs, traditional as well as modern. Since this notion of the Tao has been common since ancient China, a more democratic spirit should have been circulating in China's long history. As we all know, of course, this is not the case. This is because ancient Chinese thinkers differed from each other on the issue of the actual human potential in realizing the potential of the Tao, thus believing in different ways of achieving human potential. The predominant force of Confucianism, for example, believed in the rituals and rules set up for ordinary people who did not have the potential to comprehend the Tao if left by themselves. Only the Confucian sages were people with better potential and therefore could perceive and establish rules for those less capable masses. This emphasis on certain people's self-importance and de-emphasis on others' gave rise to elitist, meritocratic, and anti-democratic thinking.

Zhuang Zi, by contrast, successfully bypassed the trap of elitism/meritocracy without turning to the egotistical self for help. In this sense, he was ahead of his time. Even though, compared with the hereditary system, a meritocratic system was definitely a change for the better because it opened up possibilities for more people, this new system compromised the spirit of the Tao: granting some the ability to decide the Tao for others and thus violating the principle of Myriad Things. Without actually being others, no one can possibly know the myriad things in others' situations. Zhuang Zi's thought, therefore, is the most true to this principle since Zhuang Zi chooses *ziran* over the autonomous authority of sage-made social rituals.

The concept of *ziran,* however, deserves some discussion. The closest English word for *ziran* is "spontaneity," although such a translation calls for some elaboration, which I shall try to provide shortly. Clearly, the concept of spontaneity is compatible with the concept of *kairos,* the propriety of timing and measure. *Ziran* does not imply doing whatever anyone feels like doing; nor, however, does it suppress impulses and desires. *Ziran* is the Tao of impulses and desires. And this is why *ziran*/spontaneity does not mean doing nothing or being idle. Neither does it imply that the Tao is easily achieved. On the contrary, the search for the Tao involves discipline, concentration, and training. In other words, *ziran*/spontaneity is about not blowing an empty wind of words to confuse people; it means steering our desire for control based on our immediate vision of things towards spontaneous responses enabled by clear visions of myriad things. This is the reason for Zhuang Zi's belief that we cannot follow permanent rules set up by other and that we must and can make choices on our own. Zhuang Zi was fond of using examples of carpenters, sick, disabled people and animals; and Chapter II of *Zhuang Zi* starts with a person named Zi

Qi of South Wall, where common people lived. Zhuang Zi believed that carpenters or any other common people could and must be sages themselves. No one else could provide absolute rules for them; the Tao of carpentry was shaped by, as well as shaping, the myriad other things in carpentry and, therefore, could only be obtained by the carpenters themselves based on their perception and assessment of all factors involved in each given job they performed. In other words, to be sages in carpentry, they had to learn to discover the Tao themselves.

Finally, *ziran*/spontaneity also necessitates that we constantly take up new challenges. According to Zhuang Zi, the apprentice carpenters observed the Tao, but when they gained the skill, it did not mean that they no longer had to work. Once they were on the Way, they no longer had to follow the rules blindly. Without question, however, their spontaneous work, being in harmony with the Tao in each life task, was very hard work. *Ziran* meant that in the dynamic nature of the Tao, being formed and forming constantly in any given situation, successful carpenters must be forever vigilant and responsive to the situational demands on their skills.

Even though *kairos* is as difficult to translate into Chinese as into English, and even though the English word *spontaneity* cannot totally capture the phrase *ziran* the way Zhuang Zi uses it, throughout this essay I have argued that Zhuang Zi's concept of *ziran* and his understanding of the Tao are kairotic and rhetorical. Such a rhetorical view of the dynamic boundaries of differences promotes a more provisional view of truths, a more participatory discursive practice, a more tolerant cultural atmosphere, and therefore a more democratic social system. Not only can this view connect more people regardless of their professions and ages—"The Holy Man embraces them all without discriminations" (trans. Wu 146)—but also it can reach out and connect those before and those to come—"Great sages such as are met but once in ten-thousand generations are met daily by those who know how to interpret this" (trans. Cleary 79). When we learn to value the input of these "daily" sages, even if they are not college educated, then myriad things in China's democratic movement—Confucian moralism, the Declaration of Independence, the Statue of Liberty—will not appear as different as they were made to seem in 1989.

In the light of all this, the reading that Zhuang Zi is a total relativist and/or that he does not deserve serious scholarship does him injustice. To Zhuang Zi, the long and the short, the three and the four, as well as the right and the wrong have the potential to be the same because they are encompassed within the all-inclusive Tao. However, erasing apparent differences, equating the right directly and simplistically with the wrong makes not only the Tao confusing but also Zhuang Zi's whole argument meaningless. Acknowledging and sustaining, as opposed to denying, apparent differences, Zhuang Zi's real interest is in the complex nature of and relations among them. He is the one who is braving the nature of dynamics, of differences, of tough facts of life.

As I have argued, it is ironic to interpret Zhuang Zi as being anti-rhetoric simply because he says that silence can be more in tune with the Tao than words. Zhuang Zi, as shown in this essay, was against the arhetorical use of language; he was against placing formulated rules and words up and above other parts of the whole, as if their value were, in some kind of a hierarchy, above and beyond other parts of the whole, a view in essence that isolates and paralyzes language use. Zhuang Zi was true to the probable/whole nature of the Tao and always kept an eye on the kairotic/dynamic individual human being within the whole picture of myriad things. His words that still ring true in today's social-cultural events in China. Furthermore, the simplistic interpretation of Zhuang Zi does not represent a complete picture of rhetoric either, merely construing the rhetor as the conduit, not the constructor, in the rhetorical process, as if language and thinking were totally separated. In conclusion, Zhuang Zi's is indeed a much more complex and comprehensive view of discourse.

It is not whether Zhuang Zi rationalizes or uses language, but when, where, and how he does either that is the key focus of his thoughts. And it is not that Zhuang Zi sees no right or wrong; rather, he sees more clearly than many others the harm of oversimplifying the binaries of big/small, long/short, language/silence, and good/bad. Finally, for Zhuang Zi, as for many ancient Chinese thinkers, it is clear that we deal with probabilities not because we choose to, leaving certainties for some others to deal with; we deal with probabilities because that is the right way, the Way, the Tao. Given all this, even Zhuang Zi's warning against the inherent pitfalls of language use is rhetorical.

Notes

I wish to thank Roy Schoen and Janice Lauer for their help with this essay.

1. The author uses the *pin-yin* system for the Chinese names and words in her own text but keeps the authors' preferences for other phonetic systems as is in quotations.

Works Cited

Callahan, W. A. "Discourse and Perspective in Daoism: A Linguistic Interpretation of *Ziran.*" *Philosophy East and West* 39 (1989): 171–89.

Graham, A. C. *Disputers of the Tao: Philosophical Argument in Ancient China.* La Salle: Open Court, 1989.

———. "Reflections and Replies." *Chinese Texts and Philosophical Contexts: Essays Dedicated to Angus C. Graham.* Ed. Henry Rosemont Jr. La Salle: Open Court, 1991. 267–322.

Han, Minzhu, and Sheng Hua, eds. *Cries for Democracy: Writings and Speeches from the 1989 Chinese Democracy Movement.* Princeton: Princeton UP, 1990.

Hansen, Chad. *Language and Logic in Ancient China.* Ann Arbor: U of Michigan P, 1964.

Nathan, Andrew J., and Tianjian Shi. "Cultural Requisites for Democracy in China: Findings from a Survey." *Daedalus* 122 (1993): 95–123.

Oksenberg, Michel, Lawrence R. Sullivan, and Marc Lambert, eds. *Beijing Spring, 1989: Confrontation and Conflict: The Basic Documents.* New York: Sharpe, 1990.

Oliver, Robert. *Culture and Communication: The Problem of Penetrating National and Cultural Boundaries.* Springfield: Thomas, 1962.

Perry, Elizabeth J. "Intellectuals and Tiananmen: Historical Perspective on an Aborted Revolution." *The Crisis of Leninism and the Decline of the Left: The Revolutions of 1989.* Ed. Daniel Chirot. Seattle: U of Washington P, 1991. 129–46.

Rosemont, Henry, Jr. *A Chinese Mirror.* La Salle: Open Court, 1991.

Smullyan, Raymond M. *The Tao Is Silent.* New York: Harper, 1977.

Wu, Kuang-Ming. *The Butterfly as Companion: Meditations on the First Three Chapters of the* Chuang Tzu. Albany: State U of New York P, 1990.

Zhuang Zi. "Chapter II: Things, Theories—Sorting Themselves Out." Trans. Kuang-Ming Wu. *The Butterfly as Companion: Meditations on the First Three Chapters of the* Chuang Tzu. Albany: State U of New York P, 1990. 113–34.

Literacy in Athens during the Archaic Period

A Prolegomenon to Rhetorical Invention

RICHARD LEO ENOS

Very much at issue in current historical work are several deceptively simple questions: What has it meant to be a reader or a writer in, say classical Athens or medieval London?

Janice Lauer and Andrea Lunsford, "The Place of Rhetoric and Composition in Doctoral Studies" (1989)

Plato, Aristotle, and Isocrates—three of ancient Greece's most prominent thinkers on rhetoric—offer widely divergent views of the relationship between writing and rhetorical invention. Plato's *Gorgias* is one of the strongest indictments against sophistic rhetoric. Plato's criticisms in the *Gorgias* focus on what he believes to be rhetoric's flaws, that it is essentially a knack or technique that neither produces nor contributes to knowledge but rather seeks only popular agreement. Plato's criticisms were directed toward oral rhetoric, but he held the same view about written rhetoric. In his later dialogue, *Phaedrus,* Plato criticized written rhetoric, asserting that it too is only a technique or skill. In short, Plato believed that the writing instruction of sophists was not an instrument for knowledge but little more than a technical skill (*Protagoras* 236C, D). In terms important to the topic of this essay, Plato did not see writing as a heuristic that would facilitate the creation or discovery of knowledge. For Plato, writing and invention were dissociated. Unlike Plato, Aristotle did believe that writing could facilitate invention. Yet, like Plato, Aristotle also believed that the current practices of writing failed to realize writing's potential as a heuristic for enhancing complex patterns of thought and expression. In the opening passages of his *Rhetoric,* Aristotle criticizes "technographers" who taught and practiced only the surface techniques of their craft and did not understand how rhetoric could be a *techne* or art for creating rational proofs (1354a–1355). Isocrates, on the farthest end of the continuum, viewed writing as endemic to higher education. In his *Antidosis,* Isocrates expresses his belief that writing is a central part of a process of social knowledge. Isocrates believed that writing was much more than a labor skill; in fact, he believed that writing was so important that excellence in literate expression could be attained at the pinnacle of education and only with the most rigorous training of the best minds. For Isocrates, writing was inherent in rhetorical invention and essential to higher education, a view that Friedrich Solmsen has called the *ratio Isocratea* (236).

The differing views of Plato, Aristotle, and Isocrates present a very important and interesting problem about writing and its relationship with rhetorical invention. In some respects, writing was viewed as a craft that was learned for functional purposes and hence was little more than a skill. In other respects, writing was viewed as inherent in intellectual processes. These wide-ranging views of writing reveal how perspectives on writing shifted from the Archaic to the Classical Periods of Athens. In the Archaic Period, writing was seen as a labor skill. In the Classical Period, writing began to be considered more than just a skill; some rhetoricians began to link it with invention, viewing it as a heuristic that facilitated complex problem-solving in rhetoric.

This essay argues that "invention" in written rhetoric has two dimensions. The early manifestation of invention in written rhetoric, the type of writing that was produced during the Archaic Period, stressed craft techniques for recording discourse. In this sense, writing was craft. The second dimension of invention, one which began to be exhibited in the Classical Period, saw writing as a heuristic that facilitated thinking and creativity. That is, writing was an inventional tool that induced creativity and levels of complexity in thought that were by degree much more difficult to attain through methods of oral deliberation and personal interaction. Plato and Aristotle's criticisms of writing are best understood when we realize that Plato was viewing writing in the first dimension. That is, both Plato and Aristotle saw writing practiced as a recording device and not in the second dimension, that is, as a creative heuristic to facilitate thought. This paper seeks to explicate the evolution of these two dimensions of invention and, in the process, to clarify why and how writing became a manifestation of a rhetorical "art." Understanding how this evolution of thought took place will not only provide a better understanding of how writing came to be seen as a heuristic for invention but will also explain what the so-called literate revolution of Athens' Classical Period entailed. During the Classical Period writing became a feature of higher education because it was recognized as a heuristic for rhetorical invention.

Presumptions and Claims about Rhetorical Invention

Romantic notions about Athens' Classical Period and her "literate revolution" are generalizable to the point of being commonplace. Ancient Athens is characterized as the first literate community of the West. These achievements are attributed to Athens because of the wealth of writings in such arts as drama and poetry. Correspondingly, Athenian talent in non-fictive discourse is evident—also through extant writings—in philosophy, history, and rhetoric. At the height of her democracy, Athens was experiencing unparalleled intellectual achievements. One such achievement was the creation of literate rhetoric. Principles of oral rhetoric, especially the heuristics of invention, were applied to writing and the literate arts flourished. Once seen as a mundane skill, writing and rhetoric became a part of higher education, and literacy became widespread in a manner and to a degree unique to Greece. The evidence of widespread literacy in Classical Athens, occurring when rhetoric was flourishing in the school of Isocrates and through the teachings of the sophists, has prompted claims about the power of literacy. Specifically, a number of claims have been made about these social and cultural changes: the "literate" revolution of the fourth century B.C.E. changed the notion of law and justice from situational to recorded precedent (Garner); writing facilitated abstract thinking in philosophy (Havelock 1963); writing revolutionized historiography from narrative chronicling to interpretative argument of events and their meaning (Enos); and writing created a literary rhetoric through the schools such as Isocrates' (Marrou). Essentially, these claims are accurate. Law did become more systematically codified, and legal procedures did emphasize recorded precedents by the fourth century B.C.E.. A number of advancements in philosophy did occur as philosophers became literate. The flowering of historiography, begun by Herodotus and Thucydides, became inherently a writing practice by the fourth century B.C.E.. It is undeniable that these advancements occurred when writing was emphasized by sophists as a feature of higher education. It is no great proclamation to assert that the advantages of literacy became apparent when writing became rhetorical; that is, when writing shifted from a recording device to a heuristic facilitating the invention and production of discourse. What is important, and the focus of this work, is to understand the forces that contributed to this transformation. To understand this literate revolution, however, we need to understand not only literacy during the Classical Period (fifth and fourth centuries B.C.E.) but the preceding Archaic Period (late seventh to early fifth centuries B.C.E.). In short, to understand how writing and reading, tools initially used for memory and storage, became a heuristic for rhetorical invention, we need to understand the forces at work that made this transformation possible.

Current research on literacy during the Archaic Period does not clarify the complex problems of understanding writing and its relationship to invention in the Classical Period. It should be stressed at the outset that there is evidence of some literary writing in Greece as early as the seventh and sixth centuries B.C.E., but, since we possess no such manuscripts directly, we must trust later sources that book production was not in common practice until the fifth century B.C.E. (Diringer 230–33). The possible existence of recorded manuscripts, however, is not synonymous with widespread literacy or rhetorical sophistication, particularly since these writings (as will be discussed later) may have been the specialized craft of such composers as Homeric rhapsodes and other oral poets. In short, there are strong, rival positions about the nature and extent of literacy in Greece's Archaic Period, and these interpretations are no less acute when discussing Athens. Hurwit has done a masterful job of showing the depth of disagreement among scholars when discussing literacy and Athens during the Archaic Period. L. H. Jeffery, for example, argues that Greeks—especially in the latter phases of the Archaic Period—demonstrated widespread (if not sophisticated) literacy. Conversely, Eric A. Havelock, in his essay, "The Preliteracy of the Greeks," argues that the only real demonstrations of literacy during this period belonged to craftsmen and traders, a sort of "craft literacy" that will be discussed later in this essay. The incompatibility of these views is not lost upon Hurwit, who, while sympathetic to Havelock, nonetheless finds it difficult to imagine that public monuments, bearing inscriptions and built during the Archaic Period, were not intended for popular readers (193).

William V. Harris provides the best summary of the problem framed by the conflicting views of Havelock and Jeffrey. What little is actually known about literacy in the Archaic Period leads Harris to believe that by 480 B.C.E. "Greeks had attained something close to craftsman's literacy" but by no more than 10 percent of the population (59, 61). "To solve this problem," Harris states, "would require a far better knowledge of the mentality of the archaic Greeks than sources allow us" (59). The pragmatic historian of rhetoric may argue that regardless of our quibbling over whatever causal and coexistent relationships exist, the current view is clear: Athens became intellectually great when and because that community became literate. Even this view, however, is ripe for qualification, since it presupposes that there is some meaning of literacy that really is primal, generic, and univocal. The effort here is to situate the meaning of literacy by seeking to answer such questions as: "What exactly does it mean to be literate in this community?"; "Who is literate in this community?"; and "What are the social consequences of this sort of literacy?" Contextualized answers for literacy and ancient Athens tell us much about a community that had a profound historical impact. Kevin Robb makes this point about contextual perspective very well in his discussion on orality, literacy, and the alphabet in ancient Greece and the historical inferences on "widespread literacy." "Perversely," Robb argues, "as part of the same literate bias—one born of familiarity, not meanness—what we do transfer back to ancient Greece is the assumption that if anyone would read and write to any degree in any period, it would surely have to be nobles or aristocrats, or at least the wealthy" (13). Robb has captured well the point of this essay: to understand what literacy was in ancient Athens we must understand what literacy was in that community in all of its range of meanings and within specific periods if we hope to grasp the impact of rhetorical invention.

This essay seeks to resolve these and related discordant views by arguing for a reconsideration of the notions of craft and functional literacy as well as the relationship between writing and reading along class lines. Through this understanding we will be better able to realize the nature and impact that rhetorical invention had on the transformation of literacy. Such a task requires that we better understand manifestations of this Archaic mentality by re-examining our own presumptions about literacy in two ways: first, to limit the analysis to Athens; second, to examine recent archaeological evidence and current research done by scholars such as Harris and Robb. This examination of writing habits will be grounded within the context of Athens' social class structure and perspectives on orality and literacy. In short, a focus on Athenian-based evidence within the context of Athenian social history will provide a more precise framework to advance claims about literacy and its relationship to rhetorical invention than generalizations that take in all of Greece.

The Transition from Oral Rhetoric: A New Perspective toward Craft and Functional Literacy

To appreciate and understand the differences between our current view of literacy of ancient Athens and the meaning of our recent discoveries, we must first understand prevailing characterizations. Traditionally, Greek culture up through the fifth century B.C.E. has been viewed as an oral culture. As mentioned earlier, the great literate revolution has been seen as a phenomenon developing in the fifth century B.C.E. and fully manifested in the fourth century B.C.E. in the codification of law and by "writers" of theater, philosophy, history. Education itself was oral. The *symposium* was an oral-based practice where mentors, often family elders, imparted education directly to youth. In addition to being oral, Athenian education was not technical. That is, with the exception of medicine, education meant the quest for *paedeia,* or the virtue of intellectual excellence, which came about through the study of Homer, theater, music, philosophy, and oral rhetoric. The revolution in education and, it was claimed, in Greek thought itself, came about when writing was introduced. Evidence of the practical use of writing and reading has been unearthed, by relative standards; archaeological research of this century made much of this material available for study only within the last several decades. These recent archaeological findings offer new evidence about literacy in ancient Athens and its inevitable relationship with the instructional practices of that period. In contrast to the more esoteric forms of composition mentioned above, this writing is often concerned with the day-to-day activities of Athenian life, telling us much about how writing functioned in the daily practices and culture of Athens. Such evidence offers a more representative picture of what a "literate community" Athens was than our conventional resources permit. In short, we are now in a position to understand the nature of everyday literacy through evidence that was unavailable to scholars of previous centuries.

In their edited volume, *Perspectives on Literacy,* Eugene R. Kintgen, Barry Kroll, and Mike Rose illustrate the range of meaning that "literacy" can imply. Social efforts to determine the degree of literacy of a group have ranged in criteria as widely as essayist prose to the ability of an individual to demonstrate little beyond writing her or his name. Rather than to impose a standard of literacy, the effort here will be to determine what literacy meant for the Athenian community and to examine its use in every day practice, that is, the range of common social activities. Understanding the nature and functions of such writing requires that we explain some of the central terms of our discipline that describe such activities and that we modify those terms for our purpose's. Eric Havelock (*The Literate Revolution in Greece and Its Cultural Consequences*) and Walter Ong (*Orality and Literacy: The Technologizing of the Word*) have done an excellent job of explaining a range of meaning of literacy. Havelock uses the term "craft literacy" to discuss writing systems that require inordinate resources in order to function. Egyptian hieroglyphics exemplify a writing system that represents this definition of craft literacy. In order to function as a writing system, hieroglyphics required extraordinary conditions. The system of writing itself was so complex that an individual had to view the mastery of the system as entry into the profession of scribe. Such writing systems required scribes because it took a life's vocation to master the technological systems of reading and writing. To have a scribe, in turn, required resources that enabled one to afford such a system. The immense wealth of a pharaoh could provide such funding and environment. In such conditions, however, literacy was a commodity that few could afford and fewer controlled. It is obvious, therefore, that the control of such literacy did not make it public in the sense of communal use. The form systems are often viewed as less common and therefore less public; some even consider such systems as less "democratic." One speaks of such distinctions when discussing craft literacy as described by Havelock.

Craft literacy is also discussed as a way of underscoring the advantages of the alphabetic system of literacy. The alphabet, in contrast to writing systems that are descriptive-representational (hieroglyphics) and syllabaries (Linear B and Japanese) are much easier to use in many respects (e.g., *The World's Writing Systems*). An alphabet is often contrasted with a craft writing system because it can be taught and learned at a very early age; in short, no special skills or technologies are required beyond those that are readily accessible to all. Contrasted with craft literacy, the alphabetic system is much more open and public in its literacy by virtue of its ease.

The technology of the alphabetic system is also a point used to contrast its advantages with the craft system. Writing with the alphabet requires little artistic skill beyond clarity. Material is also readily available, ranging from clay (the least expensive of all writing material) to vellum and marble (the most expensive and enduring of all material). The same is true for its writing instruments, ranging from sticks to keyboard. The latter points of artistic skill, availability of material and simplicity of tools, are also available to craft systems and not really points of distinction. It is worthy of note, however, that craft systems, while they could elect to use such common material, often entail genre expectations that have high aesthetic standards and use materials that are frequently difficult and expensive to acquire. When these standards are imposed on alphabetic systems, the concept of writing as a "craft" occurs. In this respect, illuminated manuscripts, even when using an alphabet, are a form of craft literacy since they require such extraordinary talent, time, and material.

The central points of distinction commonly made between craft literacy and the alphabetic systems of composition focus on the degree of difficulty of the system and the quality and rarity of materials. These distinctions, it should be apparent, also need to be qualified, for there are occasions when alphabetic writing requires special skills or functions to the extent that it becomes a craft. As is the case with ancient Athens, it is important to recognize how and in what ways an alphabetic system of writing can be a manifestation of craft literacy. With the new perspective advocated here, we can say that craft writing systems do occur in alphabet literacy. As with Havelock, Ong also recognizes that certain forms of writing—particularly the highly complex ones—function at levels of expertise. That is, writing systems (including the alphabetic ones) can still have writing functions that require unique resources in time and talent. Just as we could say that anyone who can have access to paint, brush, and canvas can be a painter, we still recognize that highly refined and functional skills can make the term "painter" stand for a unique person who is so proficient that the activity becomes a craft. In fact, the individual may even be elevated to the distinction of an artist. These functions and levels of expertise in writing can tell us much about a group's literacy practices. In short, the "craft" of writing in nonspecialized writing systems such as the alphabet is a route toward understanding much about how literacy functions in a society. For our purposes, the task is to explicate and understand the variety of manifestations of the craft of writing at Athens, for through such an understanding we can provide much more specific knowledge about its operations and impact within the Athenian community.

Current distinctions between craft literacy and the alphabetic systems encourage scholars to advance a number of categorical inferences. Many of our scholars conclude that craft systems are technologies of writing that do not make communal literacy possible and—at least in the way that they have been characterized earlier—that inference is reasonable. These scholars also infer that alphabetic systems are, by default, not craft systems. This inference appears reasonable if we make the old distinctions and separate craft literacy and alphabetic systems. However, alphabetic systems in chirographic (handwritten) technologies can entail a large and complex scribal system, such as the one that Augustus orchestrated to help carry on the bureaucracy of the Roman Empire. While there is no argument that the ease of the task was immeasurably facilitated by the technology of the alphabet and the consistency of Latin, it does reveal a condition for, and illustrate a modest qualification of, an alphabetic *craft* literacy.

A much more important qualification, the one central to this proposal, is the meaning of craft literacy. It is this latter inference that needs to be modified if we are to have a better understanding of everyday writing in ancient Athens and the notion of a literate community. The modification of craft literacy advanced here is based not only on the degree of complexity of the writing systems, standards Havelock and Ong use in their respective treatments of the concepts, but also on performative functions. In this respect, the notion of "craft" literacy is compatible with Robb's characterization of it in *Literacy and Paideia in Ancient Greece* as the use of writing by craftsmen carrying out in their professions everyday communal functions. "For Greece," argues Robb, "the evidence for craft literacy—so different from either the scribe literacy or popular essayist literacy—comes earlier in the record and bulks larger than does any considerable body of evidence indicating that literacy had become a desirable or necessary accomplishment of the aristocratic class" (13). In fact, it is Robb's assertion that this mundane form of writing precedes the more refined writing associated with the Athenian nobility, who elected to remain traditionally athletic and oral in their education during the Archaic Period (13).

Evidence of Craft and Functional Literacy in Archaic Athens

It is not a "new" perspective to claim that literacy is contextual. That is, we determine and understand a society's uses of reading and writing by studying how literacy functions in context. If, however, we are to understand what it means to say that classical Athens was a "literate" society, we are required not only to understand the works of her luminaries, but also to take a fresh look at the mundane, commonplace practice of literacy within the culture. The first important perspective is to recognize that early literate efforts in Athens during the Archaic Period were both craft *and* functional; that is, writing was a developed trade skill and writing was used by nonexperts for facilitating everyday activities. The tasks of writing were not directed by literary artistry but rather were craft driven and task oriented. Tony M. Lentz has shown that the writing to which Socrates, Plato, and Xenophon were exposed—the sort of writing that was handed down by earlier generations—was typically done for very pragmatic, even mundane purposes: the listings of household items, records of land ownership, registers of citizens, and rolls of males eligible for service in the military (61). From our earliest manifestations of Athenian literacy during the Archaic Period it is clear that writing was a skill done for specific purposes. Moreover, this type of writing was not done by aristocrats but rather by individuals regarded as coming from lower classes.

In the Archaic Period, and into the fifth century B.C.E., writing and reading were socially divided. The tasks of writing in the earlier periods of Athenian society were done by the *thetes* or labor/craft class. In Athens, and especially after Solon's (fl. 594/593 B.C.E.) constitutional reforms, they were categorized as the lowest of the four census-classes of citizens: *pentakosiomedimnoi* (aristocrats), *hippeis* (knights), *zeugitai* (farmers), and *thetes* (laborers). *Thetes* became the craftsmen or *demiourgoi* of Athens during this period. For this group, writing was a trade skill learned for functional purposes. Epigraphical inscriptions recorded a variety of these early functions: *horos* stones (boundary markers), name-labels on pottery, engraved chronicles of civic events on *stelai* (*anagraphein*), etc. Many of these artifacts are currently on display in the Agora (e.g., *The Athenian Agora* 58–59). Such writing in the late sixth and early fifth centuries B.C.E. was done by laborers. There is no evidence of writing for any sort of literary purpose as a communal enterprise before the laws of Solon (Robb 126).

A possible exception demonstrating early literacy skill close to this time frame, and mentioned at the start of this essay, should be noted but qualified. There is a strong possibility that the recording of the works of Homer and other oral poets into written texts did take place before widespread, communal literacy. Further, there is a strong argument that a systematic demonstration of literacy took place in the sixth century B.C.E. during the rule of Pisistratus (c. 565 B.C.E.) with the recording of Homeric verse; the *Iliad* and *Odyssey* would certainly qualify as "literature." It should be understood, however, that this composition—if it did indeed happen and happen at that time—was a special and highly technical function, orchestrated under a polemarch and requiring the expertise of several rhapsodes (Enos 18–19). In short, this "Pisistratean Recension" itself is a case of craft literacy. The possibility of the profession of literary "scribe" is further evident in the statues of three writers found on the Acropolis and housed at the Acropolis Museum. As Havelock observes, these archaic sculptures offer visual representation of the special task of writing, implying a purpose and sophistication warranting their depiction with some honorific status. If these statues represent the dignification of a scribal profession, we can also infer that such proficiency in composition was considered extraordinary. It should also be realized that composition of such art was done for the purpose of public recitations and not individual, private reading. In fact, the first suggestion of "private" reading is not recorded until Aristophanes' production of the *Frogs* in 405 B.C.E. (52–67; Havelock 201–2).

Eventually, Athenian craft writing became specialized. The *hypogrammateus* (assistant secretary), for example, was in charge of recording events in the oral transactions of civic deliberations. Plato refers to these composers as the "letter-makers" who have a skill of composing functional literacy tasks (*Protagoras* 326B–D). When ostracism, the banishment by popular vote of an Athenian who had fallen out of public favor, was introduced by Cleisthenes in 508/507 B.C.E., citizens recorded their votes on potsherds or *ostraca*. Some researchers have taken this practice as a sign of widespread literacy. An examination of these *ostraca*, many of which have been unearthed in Athens with some on display in the Agora Museum (Case 38), shows little more than the ability to scratch out letters of the alphabet in a primitive way. Moreover, there is also some good reason to believe that it was not uncommon for illiterate citizens to ask fellow literate citizens to scratch down a name for them (*The Athenian Agora* 256–57; Harris 54–114; Broneer 242–43). It is obvious that writing was directed at other citizens, and we may justly assume that these higher classes of Athenian citizens were literate. It is apparent, however, that their literacy was directed toward reading and not composing, at least at this early stage. Athenians of the upper classes, in these earlier periods, were readers far more than they were composers.

The development of writing as a craft-trade, however, should not lead us to believe that other classes of Athenian citizens during the Archaic Period were nonliterate. There is evidence that other Athenian citizens did write for functional everyday purposes but not as a trade-skill nor (at this period) as a literary art. The natural inclination for historians of rhetoric is to presume that Athenians of the Archaic Period would learn to read and write at the same time (Harvey 64) and use these skills together. At the most basic level, this is likely the case. While the acquisition of reading and writing is virtually simultaneous under normal conditions, their use need not be. We know, for example, that medieval monks learned to copy manuscripts but that some of these scribes could not read what they copied (Troll). Possibly, *thetes* could not read what they wrote, but that is unlikely. The point, however, is that it was this group that engaged in the act of writing, and that act was a craft, a trade-skill. Early Athenians did learn about reading and writing but tended to separate, by degrees of emphasis, writing and reading along class lines. At least overtly, class status in Athens, since the reforms of Solon, was delineated by monetary criteria and not just family lineage. Writing was a craft, a trade to be studied and practiced, usually by citizens of the lowest class. Correspondingly, the consumers of this craft, the city and particularly the wealthy, were the beneficiary-readers.

This conditional distinction between writing and reading should not, as mentioned above, lead us to think that only the lowest classes in Archaic Athens wrote. On the contrary, we have graffiti on a range of material: pottery, shopping lists, lead tablets, bronze spear butts, and *ostraca.* Normally author-less, these items appear to have been composed by owners, most probably from a range of citizen-classes. Such writing, however, was not the stylized craft done by *thetes* but rather childlike scrawling that rarely exceeded a few words and was written on durable but scrap material for functional purposes, much the same way we today might write out grocery and laundry lists. The earliest evidence from Archaic doodlers tends to be labels of ownership (*Graffiti* #5), obscenities of word and deed (*Graffiti* #28, #30), and brief messages (*Graffiti* #22). Even writing down to the fourth century B.C.E. tends to provide lists, functions, and messages but still demonstrates no evidence of polished literacy (e.g., *Graffiti* #49). These functional, everyday practices noted, the earliest evidence of detailed and technical writing comes from laborers and was directed toward more affluent readers. In short, during the Archaic Period it must be recognized from evidence on record that writing composed for public purposes was not viewed as a literary art but a labor trade-skill engaged in as a craft for more affluent classes of readers. Samples of functional graffiti from Athens' Archaic Period—most probably composed by those who predominately were readers—demonstrate little or no sophistication in literate compositional skills (Lang 1976). The functional and craft features of literacy were much more pervasive than we can reasonably infer from our current research. That is, such claims as the one that Plato was a "literate" mind operating in an oral society are not as precise as we have initially suspected. Research on literary practices in Archaic Athens provides a more detailed explanation of the type and level of literacy that existed prior to the Classical Period. The contributions to rhetorical invention of later Athenian luminaries of arts and letters, such as Isocrates and Plato, can be better understood when we first understand how they were situated within a community that was literate in different but important ways. The remaining section of this paper is devoted to understanding the impact of these everyday writing practices of the Athenians.

Literacy, Rhetorical Invention, and Schooling

During the fifth century B.C.E., Athenian education underwent significant changes and these changes had an impact on literacy. That is, the literate upper class moved from being consumers of writing by laborers to composers themselves. This transformation came about through significant changes in education, particularly at the most advanced levels where rhetoric was taught to the prestigious of Athenian youth. This shift in the meaning of literacy was not so much that more citizens could read—although that likely did occur—but that more citizens from classes other than the *thetes* began to write. Traditional Athenian education was Homeric, where, as stated above, individuals learned through the elders of their family. Although education in ancient Athens was never municipally regulated and systematic, in the fifth century B.C.E. education began to be more communal; that is, groups of children were educated together, and much of this education began to be given not by family members but by specialists, particularly in the case of prominent families. Small children (*paidion*) still learned at home. A young boy (*pais*) went to school and began to learn reading as well as writing from the ages of seven to fourteen. From the ages of fourteen to twenty-one an adolescent male (*meirakion*) became an *ephebe* and traditionally began military training and service. This phase of education, however, was the level taken over by the sophists at the higher levels of schooling and directed more and more attention to subjects such as philosophy and rhetoric. It is these aristocratic *meirakioi* who received the attention of sophists and Isocrates, who so unsettled conservative elements of Athens by the radical practices of teaching such writing skills as rhetorical composition and logography. At all these levels of education, reading and writing increased in influence, and composition began to be seen less as a craftsman's skill and more accepted as an educational subject. As Athenians' schooling evolved in the fifth century B.C.E., the emphasis on integrating writing and reading became apparent. The grammatist taught letters to the very young, the grammarian introduced composition, and the sophists taught rhetoric at the highest level. Writing, not just its consumption but its production, had shifted from a labor skill to become a part of the *paideia* of the privileged Athenian classes.

What becomes apparent in this evolution of education in the fifth century B.C.E. is the transformation of the notion of literacy through rhetorical invention. Once seen as a craft reserved for *thetes*, writing was assimilated into the notion of *paideia* to enhance traditional instruction. The oral education of family-oriented *symposia* shifted to the schools of specialists. In the fifth century B.C.E. "literature" in terms of plays, orations, and poetry was still oral. By the fourth century B.C.E. the technology of writing was applied more and more to civic functions such as law and the recording of political deliberations. The technology of writing was also applied to the arts and used as an instrument of knowledge in the highest schools of the sophists. As Lentz argues, "the attitude of Hellenic Greeks toward writing parallels their feelings toward logographers and sophists as the use of the new technology developed beyond the archaic model of preservation" (68). Isocrates, regarded as the first real teacher of literate rhetoric, established writing as a central feature of his famous school, which attracted the main leaders of Hellas and became the cornerstone of Hellenistic education (Marrou 91; Enos 112–17).

The inscriptional record, coupled with our historical knowledge, requires us to modify the notion of literacy in Athens. In the Archaic

Period, and even into the later decades of the fifth century B.C.E., writing was a craft monopolized by the trades of the lowest class of Athenian citizens. The transformation of education, and the corresponding transformation of writing from a trade and mundane function to an educational skill, modified the concept of *paideia* and vaulted writing from the status of a craft to an art. The process of composing, assimilating principles derived from oral rhetoric, made writing more than a recording device but a system that provided the heuristics for composing literate discourses. Oral rhetoric, momentary and situational, now had a literate counterpart with a reading audience, one that was greatly dilated in time and place and freed from the constraints of spontaneous communication. Writing had evolved from the fifth and into the fourth centuries B.C.E. as a rhetorical art of invention that helped in turn to transform education. The real revolution in literacy in Athens was not so much the introduction of writing into the culture of fourth century B.C.E. Athens, but rather the transformation of writing from a craft and functional skill to an intellectual enterprise worthy of the rhetorical school of Isocrates.

Works Cited

The Athenian Agora. American School for Classical Studies at Athens. 4th ed. Princeton: Institute for Advanced Study, 1990.

Broneer, Oscar. "Excavations on the North Slope of the Acropolis, 1937." *Hesperia* 7 (1938): 161–263.

Daniels, Peter D., and William Bright, eds. *The World's Writing Systems*. New York: Oxford UP, 1996.

Diringer, David. *The Book Before Printing: Ancient, Medieval and Oriental*. New York: Dover, 1982.

Enos, Richard Leo. *Greek Rhetoric Before Aristotle*. Prospect Heights: Waveland P, 1993.

Frel, Jiri. *Panathenaic Prize Amphoras. Kerameikos Book: 2*. German Archaeological Institute Athens. Athens: Experos, 1973.

Garner, Richard. *Law and Society in Classical Athens*. New York: St. Martin's, 1987.

Harris, William V. *Ancient Literacy*. Cambridge: Harvard UP, 1989.

Harvey, David. "Greeks and Romans Learn to Write." *Communication Arts in the Ancient World*. Ed. Eric A. Havelock and Jackson P. Hershbell. Humanistic Studies in the Communication Arts. New York: Hastings House, 1978. 63–80.

Havelock, Eric. A. *The Literate Revolution in Greece and Its Cultural Consequences*. Princeton: Princeton UP, 1982.

———. *Preface to Plato.* Cambridge: Belknap-Harvard UP, 1963.

Hurwit, Jeffrey M. "The Words in the Image: Orality, Literacy, and Early Greek Art." *Word and Image* 6 (April–June 1990): 180–97.

Jeffery, L. H. *The Local Scripts of Archaic Greece.* Rev. ed. Oxford: Clarendon, 1990.

Kingten, Eugene R., Barry M. Kroll, and Mike Rose, eds. *Perspectives on Literacy.* Carbondale: Southern Illinois UP, 1988.

Lang, Mabel. *Graffiti and Dipinti.* The Athenian Agora: Results of Excavations. 21. Princeton: The American School of Classical Studies at Athens, 1976.

———. *Graffiti in the Athenian Agora.* Excavations of the Athenian Agora: Picture Books 14. Princeton: American School of Classical Studies at Athens, 1974.

———. *Socrates in the Agora.* Excavations of the Athenian Agora: Picture Books 17. Princeton: American School of Classical Studies at Athens, 1978.

Lauer, Janice M., and Andrea Lunsford. "The Place of Rhetoric and Composition in Doctoral Studies." *The Future of Doctoral Studies in English.* Ed. Andrea Lunsford, Helene Moglen, and James Slevin. New York: MLA, 1989. 106–10.

Lentz, Tony M. "Writing as Sophistry: From Preservation to Persuasion." *Quarterly Journal of Speech* 68 (1982): 60–68.

Marrou, H. I. *A History of Education in Antiquity.* Trans. George Lamb. Madison: U of Wisconsin P, 1956. Rpt. 1982.

Ong, Walter J. *Orality and Literacy: The Technologizing of the Word.* London: Methuen, 1982.

Robb, Kevin. *Literacy and Paideia in Ancient Greece.* New York: Oxford UP, 1994.

Solmsen, Friedrich. "The Aristotelian Tradition in Ancient Rhetoric." *Landmark Essays on Aristotelian Rhetoric.* Ed. Richard Leo Enos and Lois Peters Agnew. Mahway: Erlbaum, 1998.

Thompson, Edward Maude. *A Handbook of Greek and Latin Paleography.* Chicago: Ares, 1975.

Troll, Denise. "The Illiterate Mode of Written Communication: The Work of the Medieval Scribe." *Oral and Written Communication: Historical Approaches.* Written Communication Annual. 4. Ed. Richard Leo Enos. Newbury Park: Sage, 1990. 96–125.

Wycherley, R. E. *Literary and Epigraphical Testimonia.* The Athenian Agora: Results of Excavations. 3. Princeton: The American School of Classical Studies at Athens, 1957.

Vico's Triangular Invention

MARK T. WILLIAMS
THERESA ENOS

Rhetorical situations are complex sites. Wondering how these places can be described, characterized, and created, Lloyd Bitzer suggested that they are natural contexts of persons, events, objects, relations, and exigence (5). Many scholars have questioned the nature of these sites since then, and many agree that a rhetorical context is a bounded site, an area that can be defined, a scene that can be identified. Consequently, spatial analogies are frequently offered as examples of inventive processes in these settings: the reference triangle (Ogden and Richards 8); the wave, particle, and field (Young, Becker, and Pike 126); the pentad (Burke 163); the webs of intention (Flower 110); the pyramid (Bazerman 163); the map to explore rhetorical territories (Glenn 287).

Geometric figures and spatial analogies have long been used to represent how humans invent and arrange meaning. Logicians, mathematicians, and philosophers have for centuries posited the triangle to associate operations of the mind with spatial arrangements in the world (see Ong; Struever). Topics, or the places of arguments, correlate ideas in the memory with inventive strategies for particular rhetorical tasks. These topical commonplaces, what Cicero calls "regions" (*Topica* 2.7.8), enjoyed centuries of use as rhetors defined, compared, contrasted, and detailed circumstances for each particular task, particularly those related to virtue and vice. These imaginary places were turned into the Renaissance tropes of forests, gardens, theaters, and battlefields where rhetors could "hunt" for arguments and where copious invention is a process that "covers ground" (Lechner 136–37).[1]

In the early 1700s, as the powers of empirical sciences eclipsed ancient theories and practices of learning, Giambattista Vico argued for continued classroom use of the humanistic practices of rhetoric, poetics, history, and philology. A professor of rhetoric at the University of Naples, Vico stressed the importance of topics in education, the necessity of tropes in discourse, and the relevance of geometry for rhetorical invention. This paper explores Vico's triangular invention of memory, imagination, and perception—the three faculties that allow rhetors to make connections among common sense, topics, and tropes. It begins by locating Vico within rhetorical traditions and outlining his quarrel with Descartes, together with Vico's plea that rhetorical invention must be taught if students are to discover meaning in rhetorical realms. The paper examines Vico's conceptions of *ingenium,* imagination, and the rhetorical situation. The focus of this discussion, however, is Vico's understanding of the inventional function of common sense, topics, and tropes and his concept of triangular invention.

Vico in Rhetorical Traditions

Much of Vico's work is a complex etymology that reconstructs the poetic wisdom of the first humans. Recounting the parables, fables, and first agrarian laws, Vico investigated the civil histories that were constructed on the "public ground of truth" (*New Science* 351–60). Vico's major published works include *On the Most Ancient Wisdom of the Italians* (1710), *Institutes of Oratory* (1711 and 1738), *First New Science* (1725), *Autobiography* (1725), and *On the Heroic Mind* (1732). Vico's projects are interpreted in diverse ways. His theories of history, education, and language are used in communications (McLuhan), architecture (Black), myth (Mali), language-learning (Danesi), and cultural studies (Williams). Vico's circular histories —the age of gods, of heroes, of humans—and his four-stage theory of tropes have informed historiography developments in literature (Adams; Bloom) and in history (White; Kellner). Others offer links between Vico and more contemporary thinkers. In his discussion of I. A. Richards, for example, Russo calls Vico an intellectual outsider, "so far ahead of his time that he was not fully capable of being understood, let alone appreciated" (116). Bloom writes that Kenneth Burke is Vico's "true son . . . the Vico of our century" (334). Schaeffer suggests that the works of Vico and Burke are sometimes seen as "amateurish and unreadable," but these misreadings result from the fact that critics are unable to systematize their thinking (*Vico* 7). Palmer claims that Vico's criticism of Descartes compares to critiques found in the works of C. S. Peirce and John Dewey (2). Others illustrate how Vico was a reactionary, arguing as he did for continued use of humanist tradition when the new sciences gained prestige and power (Bizzell and Herzberg 647). Although Vickers blames Vico for reducing rhetoric to a study of metaphor, metonymy, synecdoche, irony (35),[2] D'Angelo credits Vico with defining relationships among tropes, *topoi,* and rhetorical invention (93). Verene asserts that Vico's concern for the "whole" and the "totality" of learning links him with Cicero (7), and Grassi connects Vico with the traditions and ideals of Italian humanism, in an intellectual line stretching from Cicero to Vives and Bruni. According to White, Vico continues the humanist connection between reason and the imagination (144).

Vico's humanist concern for a blend of imagination and reason stemmed in part from the philosophical methods of Descartes and the empirical powers of the new sciences. Descartes devalued humanist learning, which he characterized as "occupied in making the best of mere verisimilitude" (*Discourse* 54); he claimed that eloquence could be attained by speakers who are ignorant of the rules of rhetoric (7). Arnauld codified this belief in the *Port-Royal Logic,* arguing that language needed to be fixed and unambiguous.

Speaking one-half century after the publication of Arnauld's text, Vico claimed that ambiguity is unavoidable in language but that this uncertainty can be negotiated with the art of topics: "In our days, we keep away from the art of inventing arguments," he said. "We hear people affirming that, if individuals are critically endowed, it is sufficient to teach them a certain subject, and they will have the capacity to discover whether there is any truth in that subject" (*Methods* 14). Vico admits the power of "critical" or rational philosophy, but he also argues for the art of topics to discover patterns and probabilities. This debate provides a framework for Vico's use of common sense and triangular invention.

Ingenium, Invention, and the Rhetorical Situation

Researching the etymology of words in *The Most Ancient Wisdom Of the Italians,* Vico considers the term *facultas,* or faculties, which he defines as "a ready disposition for making" (93). He proceeds to classify several kinds of faculties, including imagination and memory, which can be enhanced by the study of arithmetic and geometry. These subjects of study are human faculties, "since in them we demonstrate a truth because we make it" (94). In other words, geometers can understand how their figures are created because they make them.

For Vico, this power of making has important implications for rhetoric. Describing how the rhetorical faculty of *ingenium* arises in part from a study of geometry, he writes: "*Ingenium* . . . connects disparate and diverse things. The Latins called it acute or obtuse, both terms being derived from geometry" (96–97). *Ingenium* has a long history of use and a variety of connotations, ranging from ingenuity, inventiveness, to mother wit.[3] Vico uses ingenium to draw together those things and ideas perceived to be located at a distance. Similar to Aristotle's metaphoric thinking, whereby "an acute mind will perceive relationships even in things far apart" (*Rhetoric* 1412a), Vico argues that *ingenium* can connect what was previously disconnected and that rhetors can discover new relationships in particular communicative settings. While these discoveries may not be empirically verifiable, they are true inventions because we make them. This is the *verum factum,* perhaps Vico's most frequently cited claim: "We can know the civil world because we have made it" (*Methods* 53). Vico argues that philology, history, poetics, and rhetoric are essential for study because these humanistic processes can discover how meaning was made in the formation of public institutions, as well as in fables, myths, and religion.

In *The New Science,* Vico describes these inventive faculties as *verum-factum* in this way: "[I]n the night of a thick darkness enveloping the earliest antiquity, so remote from ourselves, there shines the eternal and never failing light of a truth beyond all question: that the world of civil society has certainly been made by men, and that its principles are therefore to be found within the modifications of our own human mind" (331). Humans can discover relationships in the public world because they make these relationships with the modification of their minds, just as geometricians can know spatial figures because they construct them.[4]

Vico's references to faculties and geometry are complex, relating as they do to ancient, medieval, and Renaissance traditions (see Palmer; Verene; Goetsch). Unlike Pythagorean ideas of spatial forms as ideal representations of the physical world, Vico binds the spatial figures of geometry to the discovery of arguments in the public world. Countering Ong's suggestions that medieval quantifications of logic represent a "silent, spatial universe" (91), Vico reminds us that the acute perception that constructs geometric figures may also sharpen the rhetorical powers needed in public debate. Vico's notions also contrast with many eighteenth-century ideas of invention, whereby topics are located in individual minds (Crowley 68).

Vico admits that to compose a particular public address according to geometric methods would remove cleverness from the oration; making spatial geometric figures alone is not sufficient to inform the decisions and discoveries that may occur in practical life (*Ancient* 98–99). Unlike the analytic, Cartesian geometry that rids reason of uncertainty and doubt, Vico claims that spatial, Euclidian geometry can enhance the powers that make these discursive connections. In his *Autobiography,* Vico calls spatial geometry a "graphic art which at once invigorates memory . . . refines imagination . . . and quickens perception" (124). Rhetors can connect past experience with present apprehensions, conceive new configurations in rhetorical settings, and see previously unperceived relationships when relying on topical strategies to discover multiple points of view and to employ the dimensions of rhetorical circumstance. Or, as Vico said in a speech to the University of Naples, *ingenium* "makes present to our eyes lands that are very far away . . . unites those things that are separated . . . overcomes the inaccessible . . . discloses what is hidden" (*Humanistic* 43).

Unlike some modern and postmodern conceptions of the imagination that suggest little concern for audience, Vico incorporates the rhetorical situation in his notion of invention and imagination. Vico dedicated *The Study Methods* to Bacon, whose claim for rhetoric is "to apply and recommend the dictates of reason to imagination, in order to excite the appetite and will" (80–81). The rhetorical triangle, as posited by Ogden and Richards, assigns imaginary lines to denote the relationships among writers, readers, and topics. For Richards, the imagination is first the "production of vivid images" (*Principles* 239). So the relationships must be discovered and constructed—not with a geometer's compass but with an imagination that perceives the similar in the dissimilar and that recalls how the past can connect with the present.

For Vico, triangular invention, the interlocking faculties of imagination, perception, and memory, helps rhetors develop eloquence and discover some of the probabilities in life. He lamented the diminished abilities in his day to discover arguments and issues in the uncertain world, in rhetorical realms that are defined in and by the multiple relationships among writer, reader, and topic. This lament continues. Bailey argued in 1964 for a revival of topics, for heuristics to discover patterns in space and time (115). Hughes later called for finding the "spacious middle ground" between facts and opinions, those places that remain "still a protoplasm, not yet a shape" (158). Kinneavy claimed that participants in each age must reformulate their own topics (249).

Common Sense as Rhetorical Invention

Vico's invention relies on humanist ideas of topics and common sense, or *sensus communis,*[5] whereby the discovery of ideas occurs prior to judgment of these ideas. Following Cicero's claim that dialectic can test truth but not discover it (*De Oratore* 2.38.158), Vico offers a notion of common sense that discovers relationships within rhetorical situations before judging these associations. This common sense relies on the topics to construe rhetorical relations across diverse communicative settings. "[T]he relationship between speaker and listeners is of the essence," he writes. "In order to be sure of having touched all the soul strings of his listeners, the orator, then, should run through the complete set of the *loci* which schematize the evidence" (*Methods* 15–16).

While participating in discussions about Cartesian methods at the Academy of the Investigators in Naples, Vico concedes the methodological powers that the French philosopher unleashed (Palmer 7). He admits that these procedures made Descartes and his followers the "grand architects of this limitless fabric of the world" (*Methods* 10). Nonetheless, Vico realizes that the procedures related to Descartes's methods can prompt students prematurely into criticism—to judging issues before discovering more about the complexities and uncertainties that attend to events and problems.

Vico's *The Study Methods of Our Time,* originally composed as inaugural lectures to faculty and students at the University of Naples, compares the ways that students learn various disciplines. In each speech Vico discusses the modern disciplines of chemistry, physics, and pharmacology, attributing the methods for learning these disciplines to Descartes. Vico recalls the topical arts instead to argue that rhetorical invention should have priority over philosophical criticism, or rational judgment. Otherwise, if students judge before discovering, they are ill prepared to engage in the life of the community, to discover wise and prudent ways to speak and write about important issues (33–34).

This plea is part of Vico's common-sense philosophy, a complex tradition of thought. In his most basic definition, Vico writes that common sense is "judgment without reflection" (*Science* 142). Continuing Cicero's claim that rational thinking "contains no directions for discovering truth, but only for testing it" (*De Oratore* 2.38.158), Vico argues that rational thinking and judgment should be preceded by the study of rhetorical invention. If not, students who are encouraged to clear their minds of error and probabilities will "break into odd and arrogant behavior" as adults (*Methods* 13). Thus, young people, he writes, should be educated in common sense—a faculty that "besides being the criterion of practical judgment, is also the guiding standard of eloquence" (13). This eloquence arises when students focus on inventing arguments, when they "discern the probabilities which surround any ordinary topic" (14).

Common sense develops when students know how to use all of the topics, when they are able to perceive extemporaneously the elements of persuasion inherent in any case or question (*Methods* 15). They must investigate the *loci* of arguments that need to be considered in each particular case (*Science* 497). If a speaker visits "all the 'places' distinguished in the Topics with a critical eye. . . . topics itself will become criticism" (*Ancient* 101). Or, in Pompa's translation, "topics themselves will become critical" (73). These ideas are complemented by recent scholarship. Verene suggests that topical methods and common sense allow glimpses of how thought brings forth form (167). Gadamer states that common sense arises through living in communities (22); Covino calls common sense a "complex of shared judgments" (59); and Schaeffer claims this faculty can arise during extemporaneous oral performances when invention, figurality, and organization emerge simultaneously (*Sensus* 151).

Vico admits that topics can lead to "falsehood" because of the probabilities in any given case, but he argues that students should be taught the art of topics at an early age, so that their common sense can be strengthened and they can increase their eloquence and prudence: "Let their imagination and memory be fortified . . . let them develop skill in debating on either side of any proposed argument. Were this done, young students, I think, would become exact in science, clever in practical matters, fluent in eloquence, imaginative in understanding poetry or painting, and strong in memorizing . . . legal studies" (*Methods* 19).

Much of Vico's complaint with Cartesian methods for learning focuses on the study of geometry. For Cartesians, the spatial figures of triangles, hexagons, and parabolas are replaced by numbers. This substitution, according to Vico, makes "numb . . . obscure . . . sluggish" the powers of memory, imagination, and perception (*Autobiography* 124). Cartesian criticism further commands students to clear their minds of probabilities, to find a certainty that can stand even when beset by doubt (*Method* 9). This kind of educational training can influence students to neglect the connections that exist between seemingly diverse academic and civil topics. Vico insists that these connections can be explored by attending more to spatial geometry.[6]

Descartes lauds eloquence and poetry, but both of these arts are the gifts of nature rather than the result of study: "Those . . . who most skillfully dispose their thoughts with a view to render them clear and intelligible, are . . . wholly ignorant of the rules of rhetoric" (*Discourse* 7). For Descartes, only scientific demonstration leads to truth; the only veracity is that about which there can be no doubt: "[R]eason already convinces me that I should abstain from the belief in things which are not entirely certain and indubitable"(7). "It will be enough," Descartes maintains, "to make me reject them all if I can find in each some ground for doubt" (17).

Vico claims that Descartes's ideas rob students of the chance to develop copious methods of thinking. By focusing on methodical doubt, students are inappropriately prepared for the practice of eloquence (*Methods* 13). Vico counteracts this focus on methodical doubt by developing connections among common sense, eloquence, and topics. These connections should be developed during the first years in the classroom: "If in the age of perception, which is youth, they would devote themselves to Topics, the art of discovery that is the special privilege of the perceptive, they would then be furnished with matter in order later to form a sound opinion on it" (*Autobiography* 124). Because our acquaintance with things must come before judgment of them, Vico writes, human minds attend to topics before considering criticism: "Topics [have] the function of making minds inventive, as criticism has that of making them exact" (*Science* 498).[7] This idea has its source in Vico's notion that imagination, memory, and perception connect with "primary operations of the mind, whose regulating art is topics, just as the regulating art of the second operation of the mind is criticism; and as the latter is the art of judging, so the former is the art of inventing" (699). Vico thus offers a systematic way to invent rhetorical probabilities; Descartes offers a systematic way to doubt and do away with probabilities.

Vico's quarrel with Descartes in some ways continues today. Student writing must be clear, coherent, and organized. These writers must connect a wide range of topics from different academic disciplines, and they must posit certain claims and support their theses with demonstrable proofs. Students' essays nonetheless must emerge from a diverse mix of perceptions, memories, and imaginations that they bring to the task. Each writer relies on past experience and learning when constructing those relationships most appropriate for the particular topic. As students attempt to connect their varied academic topics and to engage issues that are discussed in wider communities, topical invention remains as necessary today as it did nearly four hundred years ago.

Vico's Topical Invention

Vico's common-sense argument with Cartesian methods of study and his reminder that the faculties that "hew out topics" also coincide with "making minds inventive" (*Science* 497, 498). These ideas have a basis in ancient and Renaissance traditions of rhetoric. Vico sees topics as an ancient faculty of knowing, a faculty informed by perception (*Ancient* 98). He argues that students, who are "in the age of perception," should concentrate on topics, the "art of discovery that is the special privilege of the perceptive" (*Autobiography* 124). In youth, when "memory is tenacious, imagination vivid, and invention quick," students can enhance their inventive powers by studying languages and geometry (*Science* 159). Vico claims that students need to perceive the connections among topics located far apart, recall how these issues may relate with previous experiences, and imagine new arrangements among these elements. Memory, perception, and imagination help them construct topical arguments and develop acute apprehensions.

In *The New Science,* Vico states that the first humans used imaginative genera to which they reduced all the species or particulars that might be located in a genus. They made the "fable fit the character and condition" (205); they applied "sensory topics" by which they poetically constructed relations between concrete individuals and species within particular genera (495). These first humans worked to "hew out topics," to place particular things in larger classificatory systems; this topical process has the "function of making minds inventive" (497, 498) and again connects Vico with classical rhetors.

Cicero equates topics with regions of the mind, as well as the areas of argument. "It is easy to find things that are hidden if the hiding places are pointed out and marked," Cicero writes. "[W]e may define a topic as the region of an argument" (*Topica* 2.7.8) Cicero's inventive "regions" are where we trace out what we hope to discover using the acuteness of intellect on which invention depends (*De Oratore* 2.8.34, 9.38). Topics help familiarize the orator with the places in the memory where she can find the arguments and recognize them instantly; a rhetor sorts knowledge into places of the mind and draws on this information when the occasion demands (Lechner 150). Vico incorporates Cicero's separate notions of *topoi,* the *loci dialectici* and the *loci rhetorici,* into the inventive art of topics. This combination, scholars observe, has considerable generative power. According to Fáj, because judgment is delayed as rhetors consider the complex and sometimes contradictory meaning of arguments, they can arrive at a more comprehensive knowledge of the subject (92). Topics are not necessarily analytic; they may also be synthetic.

Geometric implications attend to these topical processes as well. Spatial analogies for language and invention have historically been connected with topics. Discussing scholastic methods of logic that Agricola extended into topical powers, Ong claims that because our knowledge is derived through the senses, most cognition is treated by analogy with sensory cognition. In other words, language works by putting together (composing), folding up, or folding back upon (implying), setting bounds or limits (defining), and drawing a line around or sketching (division or description). These topical strategies all rely in varying degrees on the analogy between a field of intellectual activity and an area that is apprehended visually (107). How do these spatial conceptions of knowledge and language connect classical and Vichian concepts of topics?

Because of the uncertainties that frame public discourse and the actions of people, Vico argues that those who consider topical definitions, circumstances, and relationships are able to discover the probabilities that surround particular issues. This topical process encourages students not to "step rashly" into discussion when they are still in the process of discovery (*Methods* 19). Covino uses this critique to suggest that Cartesian analytics leads students to rely on narrow procedures for evaluating ideas but does not develop the ability to generate new ideas and relationships (59).

Vico's conception of the interconnectedness of imagination, perception, and memory provides a topical framework for rhetorical invention. Rhetors position individual things within wider generic categories, define things in the world by running over the commonplaces, and perceive cause-and-effect relationships (*Science* 495, 497, 498). The topics for Vico are inventive methods to discover the probabilities that attend to rhetorical situations. The processes of invention familiarize the orator with the places in the memory where she can find the arguments and recognize them instantly (Lechner 136).

Vico complements classical and Renaissance topics in ways that add generative, inventive powers to traditional topics. Similar connections exist today. Referring to the problems writers encounter constructing connections in prose, Henry James points to the need to draw, by "a geometry of [our] own," those symbolic circles in which relations do appear to stop (5). The probable dimensions that define rhetorical situations can be made visible when rhetors rely on the powers of memory, perception, and imagination. Writers recall a topic, or a commonplace; these places are perceived as starting points for the lines of argument and arrangement that can be imagined to extend outward in new figurative connections. Students imagine how to structure experience in clear paragraphs, to recall the past in relation to present purposes, and to perceive prose in coherent ways.

Imagination is also the capacity to find connections—the topical frames of definition, comparison, relationships, and circumstances that bind potentially unlimited terrains of thought. Writers rarely imagine an isolated image; they visualize relationships. Students must conceive the rhetorical dimensions that lie among themselves, readers, and topics. Vico's triangular invention can be an art of mapping some of these probable relations. By using memory, imagining new relationships, and placing these perceptions in new arrangements, Vico's triangular invention adds common-sense dimensions to traditional topics.

Basic to these topical processes are the powers of metaphoric thinking. Near the start of *The New Science,* where Vico sets forth the principles of his text, he writes that any philological study must begin where its topic begins. He imaginatively connects these diverse times and places to offer his perception of the past: "We must . . . fetch [the topics] from the stones of Deucalion and Pyrrha, from the rocks of Amphion, from the men who sprang from the furrows of Cadmus or the hard oak of Vergil" (338). This topical etymology discovers how humans may have related diverse ideas and developed the acute wit that characterizes inventive minds. Vico's treatment of metaphor and other tropes are basic to these discoveries of meaning.

Vico's Tropes as Topical Heuristics

Vico demonstrates rhetorical invention by connecting what was previously disconnected, by uniting what was previously disunited. Vico recalls the past to perceive new links between disparate and diverse things, which he places together with imaginative, troped examples: "The great fragments of antiquity, hitherto useless to science because they lay begrimed, broken, and scattered, shed great light when cleaned, pieced together, and restored" (*Science* 357).

Vico shows how the earliest humans situated particular sensations in more general matrices of thought. The first humans needed to create classificatory concepts for the particular events and things they observed in the world. They used "imaginative class concepts . . . to reduce all the particular species which resembled them" (*Science* 209). These new turns and changing arrangements are found with the use of figurative language, what White calls the categories of Vico's poetic logic (*Science* 207). Tropes are the first references to the world and to other persons in the world, Vico writes; and these semantic turnings are important human inventions. Metaphor "gives sense and passion to insensate things" (*Science* 404) and allows the unfamiliar to be perceived in terms of the familiar. Metaphor "plays the first role in acute, figurative expression" (*Methods* 24).

According to Vico, ancient people used topical powers to relate inanimate things in the world to the human senses and passions through the powers of tropes, from the ability to "hew out topics" (*Science* 405, 495). The first humans assigned meaning to the powers of nature, such as lightning, which they imaginatively called Jove (*Science* 379). They associated unknown things in nature to parts of the human body: the human head for the top of a thing or the beginning of a process; the mouth to openings perceived in nature; the heart as the center of issues and events. They imagined the unknown in terms of the known, perceiving relations of species and genus, container and contained. The first poets used these categories to ascribe names to things, to label the most particular and sensible ideas. Such processes are the sources of metaphor, synecdoche, and metonymy (*Science* 406). These tropes are not simple surface embellishment, but are the "necessary modes of expression" that signify abstract forms or genera that contain their species or relate parts with wholes (*Science* 409). When tropes are understood as necessary modes of expression, and not as ingenious inventions of writers, two historical misconceptions of grammarians are corrected: the idea that prose speech is proper, poetry improper and the notion that prose discourse develops historically before poetic speech (*Science* 460).

This "poetic logic" influences rhetorical invention. Discovery relies on the power to relate the known to the unknown, to connect familiar images with unfamiliar images, to position parts of an issue within wider realms of meaning. These tropical processes have topical implications because metaphor suggests analogical relations, metonymy points to the arrangements of circumstances (contiguity as well as cause-and-effect relationships), and synecdoche provides ways to express part-to-whole relationships. Triangular inventions can be considered tropical discoveries because the points of a triangle or cube are the places where a pencil must turn to connect imaginary lines. Of course, rhetorical situations are immensely more complex than any spatial figure, populated as they are by people who act and interpret in multiple ways that cannot be reduced to any simple analogy or spatial figure. Still, rhetorical situations may be described in the basic topical terms of definition, circumstance, and relationship and considered in terms of parts and wholes, container and contained—as reductions of the complex to the less complex. The multiple dimensions of all rhetorical situations are thus reduced to relatively static realms on the page.

Triangular invention is another way to think about the original nature of language as metaphoric. These spatial relationships can be constructed with memories, perceptions, and imaginations to discover connections between diverse topics.

Triangular Invention

As the preceding shows, Vico relied on classical and humanist traditions to formulate his ideas on topics, tropes, and common sense as alternatives to Cartesian criticism and judgment. Rather than relying on certainty and doubt, Vico argued that the triangular faculties of memory, imagination, and perception can assist the discovery of probable relationships that comprise rhetorical situations. Vico also drew on the classical rhetorical tradition in defining his notion of triangular invention

Classical rhetoricians considered memory a region where sense perception is stored. Vico writes, "But memory also signified the faculty that fashions images. . . . Therefore, the Greeks have handed down in their myths the tradition that the Muses, forms of imagination, were the daughters of Memory" (*Ancient* 95–96). Memory makes imaginary places in the mind where invention can begin. Recalling Simonides' anecdote to illustrate the rhetorical art of memory, Cicero offers examples of the interconnected powers of memory, perception, and imagination. The most coherent mental pictures are formed in the mind by images conveyed by sight. Ideas that arrive without visual form, notions conducted through speech and reflection, will gain additional coherence with the application of visual images. The powers of memory can transform these nonvisual conceptions, ascribing to them "a sort of outline and image and shape so that we keep hold of as it were by an act of sight that we can scarcely embrace by an act of thought" (*De Oratore* 2.87.357). Situating these powers in places of the mind, Cicero suggests that an eloquent use of topics relies on powerful recall. This memory in turn relies on situated powers of perception and imagination: "we may grasp ideas by means of images and their order by means of localities" (2.88.359). Quintilian offers similar interconnected notions of memory, imagination, and perception; he suggests that the powers of memory are strengthened "by stamping localities on the mind" (*Institutes* 11.2). These localities can later be revisited with recall, and the stored images are perceived in interconnected associations: "symbols and objects recalls all of the details, for they are linked together like dancers holding hands" (11.2).

Vico relies on these classical authors throughout his work, pointing specifically to memory as he lays out his clearest depiction of triangular invention near the end of *The New Science:* "Memory thus has three different aspects: memory when it remembers things, imagination when it alters or imitates them, and invention when it gives them a new turn or puts them into proper arrangement and relationship" (819). Rhetors bring memory to every communicative event, and these memories provide a framework, or a place, where perception and imagination can arrange recollection in new relationships with changing rhetorical purposes. In other words, Vico's triangular invention occurs when rhetors can connect what might seem disconnected by bringing the three faculties of memory, imagination, and perception into a collaboration with the persons and purposes that are at hand.

Vico's triangular invention reminds us to focus on the powers of perception, memory, and imagination when we try to discover the topical relationships that determine some of the outlines of rhetorical tasks. Inventive persons, Vico writes, "are the ones who are able to find a likeness or ratio between things very different and far removed from one another, some way in which they are cognate, or who leap over the obvious and recall from distant places the connections appropriate for the things under discussion" (*Ancient* 102). Rhetors can never fully know how perceptions emerge with memory, nor can they understand exactly how imaginations inform communicative settings. Still, as writers imagine how to situate intentions in particular rhetorical situations, they consider how memory might influence their perceptions of these communicative environments. When looking for the shapes and outlines of arguments and ideas that may be appropriate for a particular rhetorical task, they can recall from distant places those topics that might be most apt. This three-stage inventiveness is a recursive way to find and arrange lines of argument and points of perception that are informed by the backgrounds of memory.

Vico links the apprehensions discovered in part through the study of spatial figures to his notion of the constructed character this of the civil world. When these geometric figures are composed for diverse, different, and complicated problems, Vico writes, they "strengthen the imagination, which is the eye of mother wit" (*Ancient* 104). These processes lead to a synthesizing power that composes meaning and that connects the familiar with the unfamiliar. Rhetors imagine things by alteration and imitation, and then invent new relationships with alternative turns of thought and speech.

Geometric figures represent three-dimensional objects on a two-dimensional surface. Triangles, cubes, and hexagons hint at the complexities in experience that can be reduced to paragraphs on the page. They are quantified analogs for the qualitative experience writers place into print. This acute inventive ability, while perhaps not improved by drawing triangles or hexagons, might be enhanced when writers remember that the relationships that comprise rhetorical settings might be perceived in new ways; the connections among readers, texts, and topics might be imagined in arrangements that better connect diverse purposes with diverse readers.

Implications

Vico descends into collective memory, the past, to perceive how language developed. Vico's triangular invention may assist the probable tasks that writers engage today, as they try to discover how the past relates to the present, how futures are informed by the histories they bring to each moment. Many memories, perceptions, and imaginations emerge in these multiple sites. Writers need to imagine how their intentions may connect, disconnect, or conflict with the purposes of others, to perceive how they might use and discover new means of persuasion, and to remember the *topoi* that will help outline the multiple dimensions of diverse rhetorical situations. Triangular invention is a way to measure place, figure space, and articulate intent.

The image of triangular invention may also help writers amplify their interpretations with turns of phrase, invigorate their apprehensions with memory, and alter their perceptions to create the probable places between books and lives. Topics may become lines of thought that connect rather than separate writers from difficult issues. Each writer must discover new regions, new pathways, as they recall experiences, imagine new connections with the present, and perceive alternative sequences in the relationships that comprise meaning. These relationships gain shape with the poetic logic that drives figurative language. When students join judgment with discovery by relying on topics, they find ways to invent meaning in the realm of the probable, in common-sense constructions discovered among the diverse relationships of rhetorical situations.

Notes

1. Lechner writes that there were at least twenty definitions of commonplaces in Renaissance thought. These definitions were based on the imaginary concepts of the writer's mind (136).

2. "No rhetorician before Vico could have thought of describing the evolution of human consciousness in terms of the interactions of four tropes" (Vickers 35).

3. See, for example, Cicero's *De Inventione* 1.30.49 and Quintillian *Institutes* 10.1.130. Grassi claims that powers of *ingenium* allowed humanist translators to reconstruct with tropes and metaphors the historical conditions of classical texts (*Renaissance* 22–25). Schaeffer claims this force construes active perception (*Sensus* 69); Verene suggests *ingenium* combines past acts with present acts (*Vico's Science of Imagination* 105); Milbank claims ingenium makes connections in the "'angular' world of geometry" as well as in the construction of meaning in rhetoric and poetry (271).

4. Some philosophers of science cite the inventive, heuristic powers of Euclidian geometry to suggest connections between how spatial figures are made and how theories are formulated (Sklar 14). Covino notes that geometry illustrates logical relationships (61), and Kunze writes that the geometric method was discovered "on a topical inventiveness for which it could not account" (87).

5. Schaeffer details how *sensus communis* has three main connotations: Plato's doxa of hearsay and opinion; Aristotle's technical definition, whereby objects are perceived by the senses simultaneously, then compared and appreciated in their differences (*De Anima* 426b8–27a15); the Roman ideal of a sense of propriety—the common but unstated mores of a community. Schaeffer argues that Vico synthesizes these notions of common sense (*Sensus Communis* 2–3).

6. In his preface to *The Study Methods,* Verene suggests that arguments of the moderns, or the Cartesians, should be studied, "but only by mature minds that have been educated in the art of topics" (xiii). Corsano suggests that some of the most important discoveries of the Renaissance in fact occurred without the use of Cartesian, analytic geometry (430). Descartes's analytic geometry nonetheless "outstripped" the synthetic geometry that Vico claimed was beneficial for students (Belaval 81).

7. Grassi writes that rhetorical discoveries cannot occur within a deductive process: "[T]he original premises as such are nondeductible and the rational process hence cannot 'find' them; . . . moreover, rational knowledge cannot be a determining factor for rhetorical or poetic speech because it cannot comprehend the particular, the individual, i.e., the concrete situation; and since the critical method always starts with a premise, its final conclusions are necessarily valid only generally" (*Rhetoric* 44).

Works Cited

Adams, Hazard. *Philosophy of the Literary Symbolic.* Tallahassee: Florida State UP, 1983.

Aristotle. *The Rhetoric.* Trans. W. Rhys Roberts. New York: McGraw, 1954.

Bacon, Francis. *The Works of Francis Bacon.* Ed. J. Spedding, R. L. Lewis, and D. D. Heath. New York: Modern Library, 1955.

Bailey, Dudley. "A Plea for a Modern Set of Topoi." *College English* 26 (1964): 111–17.

Belaval, Yvon. "Vico and Anti-Cartesianism." *Giambattista Vico: An International Symposium.* Ed. Giorgio Tagliacozzo and Hayden V. White. Baltimore: Johns Hopkins UP, 1969. 77–91.

Berlin, Isaiah. *Vico and Herder: Two Studies in the History of Ideas.* New York: Viking, 1976.

Bitzer, Lloyd. "The Rhetorical Situation." *Philosophy and Rhetoric* 1 (1968): 1–14.

Bizzell, Patricia, and Bruce Herzberg. *The Rhetorical Tradition: Readings from the Classical Times to the Present.* New York: St Martin's P, 1990.

Bloom, Harold. "Poetry, Revisionism, Repression." *Critical Theory Since 1965.* Ed. Hazard Adams and Leroy Searl. Tallahassee: Florida State UP, 1986. 331–43.

Burke, Kenneth. *A Grammar of Motives.* New York: Prentice, 1945.

Cicero. *De Oratore.* Trans. E. W. Sutton. 2 Vol. London: Heinemann, 1942.

———. *Topica.* Trans. H. M. Hubbell. New York: Loeb, 1960.

Corsano, Antonio. "Vico and Mathematics." *Giambattista Vico: An International Symposium.* Ed. Giorgio Tagliacozzo and Hayden V. White. Baltimore: Johns Hopkins UP, 1969. 425–37.

Covino, William A. *The Art of Wondering: A Revisionist Return to the History of Rhetoric.* Portsmouth: Boynton/Cook, 1988.

Croce, Bendito. *History of the Kingdom of Naples.* Trans. Frances Frenaye. Chicago: U of Chicago P, 1970.

Crowley, Sharon. *The Methodical Memory: Invention in Current Traditional Rhetoric.* Carbondale: Southern Illinois UP, 1990.

Danesi, Marcel. *Vico, Metaphor, and the Origin of Language.* Bloomington: Indiana UP, 1993.

D'Angelo, Frank J. "The Four Master Tropes: Analogues of Development." *Rhetoric Review* 11 (1992): 91–109.

Descartes, René. *A Discourse on Method.* Trans. John Veitch. New York: E. P. Dutton, 1949.

Fáj, Attila. "Vico as Philosopher of Metabasis." *Giambattista Vico's Science of Humanity.* Ed. Giorgio Tagliacozzo and Donald Phillip Verene. Baltimore: Johns Hopkins UP, 1976. 87–109.

Flower, Linda. *The Construction of Negotiated Meaning: A Social Cognitive Theory of Writing.* Carbondale: Southern Illinois UP, 1994.

Gadamer, Hans-Georg. *Truth and Method.* New York: Seabury, 1975.

Glenn, Cheryl. "Remapping Rhetorical Territory." *Rhetoric Review* 13 (1995): 287–303.

Grassi, Ernesto. *Philosophy and Rhetoric: The Humanist Tradition.* University Park: Pennsylvania State UP, 1980.

———. *Renaissance Humanism: Studies in Philosophy and Poetics.* Trans. Walter F. Veit. Binghamton: Medieval & Renaissance Texts and Studies, 1988.

Goetsch, James Robert, Jr. *Vico's Axioms: The Geometry of the Human World.* New Haven: Yale UP, 1995.

Howell, Wilbur Samuel. *Eighteenth Century British Logic and Rhetoric.* Princeton: Princeton UP, 1971.

Hughes, Richard E. "The Contemporaneity of Classical Rhetoric." *CCC* 16 (1965): 157–59.

James, Henry. "Preface to Roderick Hudson." *The Art of Fiction.* New York: Scribners, 1934. 3–19.

Kellner, Hans. *Language and Historical Representation.* Madison: U of Wisconsin P, 1989.

Kinneavy, James. *A Theory of Discourse: The Aims of Discourse.* Englewood Cliffs: Prentice, 1971.

Kunze, Donald. *Thought and Place: The Architecture of Eternal Places in the Philosophy of Giambattista Vico.* New York: Peter Lang, 1987.

Lechner, Joan Marie. *Renaissance Concepts of the Commonplaces.* New York: Pageant, 1962.

McLuhan, Marshall, and Eric McLuhan. *Laws of Media: The New Science.* Toronto: U of Toronto P, 1988.

Milbank, John. *The Religious Dimension in the Thought of Giambattista Vico, 1668-1744: Vol. I The Early Metaphysics.* Lewiston: Edwin Mellen, 1991.

Ogden, C. K., and I. A. Richards. *The Meaning of Meaning: A Study of the Influence of Language Upon Thought and of the Science of Symbolism.* New York: Harvest, 1923.

Ong, Walter J. *Ramus, Method, and the Decay of Dialogue.* Cambridge: Harvard UP, 1958.

Palmer, L. M. Introduction. *On the Most Ancient Wisdom of the Italians.* By Giambattista Vico. Ithaca: Cornell UP, 1988. 1–40.

Pompa, Leon, ed. *Vico: Selected Writings.* Cambridge: Cambridge UP, 1982.

Quintilian, Marcus Fabius. *The Institutio Oratoria.* Vol. 1. Trans. Charles Edgar Little. Nashville: George Peabody College for Teachers, 1951. 2 vols.

Richards, I. A. *Principles of Literary Criticism.* Harcourt: New York, 1985.

Russo, John Paul. *I. A. Richards: His Life and Work.* Baltimore: Johns Hopkins UP, 1989.

Schaeffer, John D. *Sensus Communis: Vico, Rhetoric, and the Limits of Relativism.* Durham: Duke UP, 1990.

———. "Vico and Kenneth Burke." *Rhetoric Society Quarterly* 26.2 (1996): 7–17.

Sklar, Lawrence. *Space, Time, and Spacetime.* Berkeley: U of California P, 1974.

Struever, Nancy S. "Vico, Valla, And the Logic of Humanist Inquiry." *Giambattista Vico's Science of Humanity.* Ed. Giorgio Tagliacozzo and Donald Phillip Verene. Baltimore: Johns Hopkins UP, 1976. 173–86.

———. *The Language of History in the Renaissance.* Princeton: Princeton UP, 1970.

Verene, Donald Phillip. Introduction. *On Humanistic Education.* By Giambattista Vico. Trans. Giorgio A. Pinton and Arthur W. Shippee. Ithaca: Cornell UP, 1993. 1–30.

———. Preface. *On the Study Methods of Our Time.* By Giambattista Vico. Trans. Elio Gianturco. Ithaca: Cornell UP, 1990. ix–xix.

———. *Vico's Science of Imagination.* Ithaca: Cornell UP, 1981.

Vickers, Brian. "The Atrophy of Modern Rhetoric: Vico to de Man." *Rhetorica* 6 (1988): 21–56.

Vico, Giambattista. *On the Most Ancient Wisdom of the Italians.* Trans. L. M. Palmer. Ithaca: Cornell UP, 1988.

———. *The Autobiography of Giambattista Vico.* Trans. Max Harold Fisch and Thomas Goddard Bergin. Ithaca: Cornell UP, 1944.

———. *On Humanistic Education.* Trans. Giorgio Pinton and Arthur W. Shippee. Ithaca: Cornell UP, 1993.

———. *The New Science.* Trans. Thomas Bergin and Max Fisch. Ithaca: Cornell UP, 1970.

———. *On the Study Methods of Our Time.* Trans. Elio Gianturco. Ithaca: Cornell UP, 1990.

Wallace, Karl R. "Topoi and the Problem of Invention." *Quarterly Journal of Speech* 58 (1972): 387–96.

White, Hayden. *Tropics of Discourse: Essays in Cultural Criticism.* Baltimore: Johns Hopkins UP, 1978.

Williams, Raymond. *The Sociology of Culture.* Chicago: U of Chicago P, 1981.

Yates, Frances A. *The Art of Memory.* London: Routledge, 1966.

Young, Richard E., Alton E. Becker, and Kenneth L. Pike. *Rhetoric: Discovery and Change.* New York: Harcourt, 1970.

Contributors

FREDERICK J. ANTCZAK is Professor of Rhetoric and Professor in the Project on Rhetoric of Inquiry, and Associate Dean for Academic Programs in the College of Liberal Arts and Sciences at the University of Iowa. Winner of a Phi Beta Kappa book award for *Thought and Character: The Rhetoric of Democratic Education,* he served as president of the Rhetoric Society of America from 2000–2001. His most recent project was editing *Professing Rhetoric: Proceedings of the 2000 Rhetoric Society of America Conference.*

JANET M. ATWILL is Associate Professor of English at the University of Tennessee, where she teaches graduate and undergraduate courses in the history of rhetoric, rhetorical and critical theory, political rhetoric, writing, and cultural studies. She is the author of *Rhetoric Reclaimed: Aristotle and the Liberal Arts Tradition* (1998) and coauthor of *Four Worlds of Writing: Inquiry and Action in Context* (2000) and *Writing: A College Handbook* (2001). She is presently completing with Donald Lazere a book entitled *Introduction to Political Rhetoric: A Guide to Civic Literacy* and collaborating with Arthur Walzer and Richard Graff on a collection of essays entitled *The Viability of the Rhetorical Tradition.* She has published articles and book chapters on Aristotle, classical rhetoric, and historiography. Her current work focuses on theories of consensus and dissent in civic and social movement rhetoric.

JULIA DEEMS is a Ph.D. candidate in English at Carnegie Mellon University. Her interests include community literacy, composition, and computer technologies to support writing and critical thinking. In her dissertation work, she developed a hypermedia program that implements heuristic support for students to think through, write about, and work to solve community problems.

Richard Leo Enos is Professor of English and Holder of the Lillian B. Radford Chair of Rhetoric and Composition at Texas Christian University. His research emphasis is in the history of rhetoric with a specialization in classical rhetoric. Much of his work deals with understanding the relationship between thought and expression in Antiquity. He has studied in Italy and Greece and has done research under the auspices of the American School of Classical Studies at Athens and the Greek Ministry of Science and Culture. He has received support for the study of ancient rhetoric from the National Endowment for the Humanities and is the recipient of the Karl R. Wallace Award and the Richard E. Young Award for research in classical rhetoric. He is a past president of the Rhetoric Society of America and has served on the Honors and Awards Committee for the Modern Language Association. He is the current editor of *Advances in the History of Rhetoric*, an annual publication of The American Society for the History of Rhetoric.

Theresa Enos is Professor of English and Director of the Rhetoric, Composition, and the Teaching of English Graduate program at the University of Arizona. Founder and editor of *Rhetoric Review,* she teaches both graduate and undergraduate courses in writing and rhetoric. Her research interests include the history and theory of rhetoric and the intellectual work and politics of rhetoric and composition studies. She has edited or coedited nine books, including *The Encyclopedia of Rhetoric and Composition, The Spaciousness of Rhetoric,* and *The Writing Program Administrator's Resource: A Guide to Reflective Institutional Practice,* and she has published numerous chapters and articles on rhetorical theory and issues in writing. She is the author of *Gender Roles and Faculty Lives in Rhetoric and Composition* (1996) and a past president of the national Council of Writing Program Administrators (1997–99).

Linda Flower is Professor of Rhetoric and Director, Center for University Outreach at Carnegie Mellon. A student of rhetoric and cognition (*The Construction of Negotiated Meaning: A Social Cognitive Theory of Writing*), her work in community literacy brings college students and urban youth into collaborative inquiries and public dialogues around urban issues (*Problem-Solving Strategies for Writing in College and Community*). Her recent book, *Learning to Rival: A Literate Practice for Intercultural Inquiry,* argues for a Deweyan, rival hypothesis stance to cultural cross-walking. It compares the learning stories and strategies of minority students, asked to engage in such inquiry, in a college class and a community center. Her current work demonstrates how a rhetorical process of intercultural (and cross-hierarchical) problem solving can bring new knowledge to practical workforce and workplace problems (The Carnegie Mellon Community Think Tank, www.cmu.edu/outreach).

DEBRA HAWHEE is Assistant Professor of English at the University of Illinois, Urbana-Champaign, where she teaches courses in rhetorical studies, body studies, and writing. She has published articles in *College Composition and Communication, Quarterly Journal of Speech, JAC: A Journal of Composition Theory,* and *Journal of Sport and Social Theory.* Her interests include classical rhetoric, body studies, and sports studies. She is presently completing a book entitled *Bodily Arts: Rhetoric and Athletics in Antiquity.*

JANICE M. LAUER is the Reece McGee Distinguished Professor of English at Purdue University, where she founded, directed, and teaches in the graduate program in Rhetoric and Composition. She received the 1998 CCCC Exemplar Award, has coauthored *Four Worlds of Writing* and *Composition Research: Empirical Designs,* coedited the composition entries in *Encyclopedia of English Studies and Language Arts*, and published on invention, persuasive writing, classical rhetoric, and composition studies as a discipline. For thirteen years she directed a national summer Rhetoric Seminar, has chaired the College Section of NCTE, and served on the executive committees of CCCC, the MLA Group on the History and Theory of Rhetoric, and the Rhetoric Society of America

DONALD LAZERE is Professor Emeritus of English at California Polytechnic State University, San Luis Obispo. He is the author of *The Unique Creation of Albert Camus* and editor of *American Media and Mass Culture: Left Perspectives*. His articles have appeared in *College English, College Composition and Communication, New Literary History, American Quarterly, The Journal of Communication, Tikkun, The Chronicle of Higher Education, The New York Times,* and *The Los Angeles Times.*

YAMENG LIU is Associate Professor of Rhetoric and English at Carnegie Mellon University. He has coedited *Landmark Essays on Rhetorical Invention,* published articles in leading national and international journals of rhetorical studies, and contributed chapters to numerous recent books on rhetorical theory and pedagogy. He is currently working on a book analyzing and critiquing the rhetoric of comparative studies.

ARABELLA LYON is Associate Professor of English at State University of New York at Buffalo. She is the author of *Intentions: Negotiated, Contested, and Ignored* (1998), a W. Ross Winterowd Award winner. Her articles on rhetorical theory have appeared in *College English, JAC: A Journal of*

Composition Theory, College Composition and Communication, and *Rhetoric Review*. She is currently working on a theory of rhetorical deliberation.

LOUISE WETHERBEE PHELPS is Professor of Writing and Rhetoric at Syracuse University, where she founded and directed the Syracuse Writing Program and initiated a doctoral program in Composition and Cultural Rhetoric. In 1993–94 she held an ACE Fellowship to study higher education and academic leadership. She is the author of *Composition as a Human Science* (1988) and coeditor of a research volume, *Feminine Principles and Women's Experience in American Composition and Rhetoric* (1995) and a teaching anthology, *Composition in Four Keys: Inquiring into the Field* (1995). She has published articles and book chapters on numerous topics in composition and administration. Currently she is completing *Poetics of Composition: Footprints of an Intellectual Journey,* the first of a two-volume collection of her published and unpublished writings.

JAY SATTERFIELD is the Reader Services Librarian in the Department of Special Collections at the University of Chicago Library. His specialty is the interplay between taste and publishing, specifically in twentieth-century America book culture. He is the author of *"The World's Best Books": Taste, Culture, and the Modern Library,* published in the University of Massachusetts Press *Studies in Print Culture and the History of the Book* series in 2002.

HAIXIA WANG is Assistant Professor in the English Department at the University of Wisconsin–La Crosse. She teaches Rhetoric/Composition and is currently Director of the Writing Center on campus. She studies cross-cultural rhetorical theory and practice, particularly between China and the West, focusing on the ways language and thought interact with the teaching of language arts.

MARK T. WILLIAMS is an Assistant Professor of English at California State University, Long Beach, where he teaches courses in composition and rhetoric and codirects the undergraduate writing program. A former journalist and middle school teacher, he has presented papers on histories of rhetoric and writing across the curriculum at Rhetoric Society of America, CCCCs, and WPA. His present work applies Kenneth Burke's Terms for Order to the invention and ordering of rhetorical contexts.

Index

Perspectives on Rhetorical Invention was designed and typeset on a Macintosh computer system using QuarkXPress software. The text is set in Caslon 224 Book and the chapter openings are set in Britannica Light. This book was designed and typeset by Bill Adams and manufactured by Thomson-Shore, Inc.